The Tale of
GOD &
CHANCE

MARK REES - THOMAS

For those I've met along the way
who know there's something more.

ACKNOWLEDGEMENTS

My sincere thanks to my wife Janine, and my children and family who hold fervently to the truth of Jesus Christ, and who inspire me to do likewise.

Most of all, I thank God who prompted me many years ago to question and examine the evidence for God, and who helped me form a firm conclusion.

ACKNOWLEDGEMENTS

CONTENTS

INTRODUCTION

At the end of the movie 'Saving Private Ryan', now an old man, Private Ryan kneels down at the graves of his Army captain and fellow soldiers. He survived while they hadn't. They died young while he lived a full life. Turning to his wife with tears in his eyes he says *"Tell me that I lived a good life. Tell me that I'm a good man"*.

Spurred on by their sacrifice, Private Ryan did his best to live with a sense of meaning and purpose.

Purpose gets us up in the morning. Even when we endure hardship and loss, we can still discover purpose each day in what we do, the lives we impact, and the legacy we build. Private Ryan carried the sorrow and memories of his fallen brothers, but still found purpose in starting a family and living a full life.

The problem however, is this type of purpose ends when we end. Sure, at our funeral, people will say nice things and recount the things we did and the love we showed, but is that really the end of the matter? Is that the extent of what our lives amount to?

When it comes to purpose, I'd be surprised if we haven't all at some point stepped back from what we do each day and asked the bigger questions *'Is there more to life than this?'*, *'What's really my purpose?'* *'Is there something else beyond this life?'*

In approximately 1000BC, the great King Solomon wrote these words *"God has set eternity in the hearts of men"*. We cannot get past the sense we have at times, that there is a greater purpose to our existence. Answering that question has been the quest of billions of people for thousands of years. It has filled libraries, philosophy classes, and cathedrals.

Evolutionary scientists answer this question by saying we are the product of chance and randomness. We are in a fight for survival and there is no greater purpose to our existence, so we must make the most of what we have. However, as we later explore, not only is that a depressing view of life, but more significantly, that explanation doesn't stack up.

Unguided evolution as an explanation for life is an unfortunate tale that has crept into our classrooms, and it runs contrary to the abundance of evidence around us. It removes purpose and hope from our young people who desperately need to hear they are valued and that there is meaning to their lives.

The truth is you are here for a purpose. In fact, you have immense purpose. You won't find it in a laboratory, but you will find it when you open your mind and heart to the question of God. Having done my homework, I have concluded the Bible is the most powerful object the world possesses because it is a personal message from the God who made us, and it answers the question of purpose through inviting us into a relationship with God. Don't worry, I'm not asking you to take my word for that claim. I discovered some time ago that the Bible can be put under a microscope, and it is up for the challenge.

Over the years I have spoken with friends, colleagues and strangers and enjoyed great conversation with them about purpose, and the existence of God. We have discussed whether there is validity to the claim that Jesus Christ is God, that he died for our sins and rose to life and that we find eternal purpose when we get to know him.

I love sitting around the outside fire at our house and sharing thoughts on a range of topics like these. I'd love to do that with you, so consider this book as one of those conversations where we journey together on the big questions of life, and as we do, I encourage you to come to your own conclusion.

———◆•◆———

THE QUESTION OF GOD

"If there is no god, nothing matters. If there is a god, nothing else matters" (H.G. Wells)

In 1968, the Apollo 8 crew spent Christmas Eve orbiting the moon. As the first manned mission to orbit the moon, they watched the beautiful blue planet earth rise above the moon's horizon. Beholding this unique perspective, they broadcast a message to earth and read the first words of the Bible *"In the beginning, God created the heavens and the earth. And the earth was without form and void, and darkness was upon the face of the deep. And the Spirit of God moved upon the face of the waters. And God said, 'let there be light'".* In the midst of mankind's remarkable achievement, those words reminded us that humanity is part of something far greater, far grander.

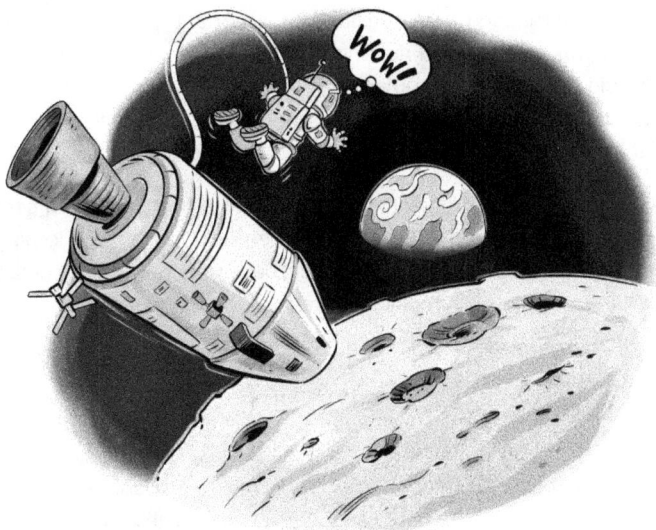

Was there a God listening into the broadcast that day, noticing the little tin can as it carried 3 astronauts further than ever before? Or were those words merely poetry reflecting man's effort to understand the universe and make sense of their existence. Either answer has profound implications for us all.

As remarkable as landing on the moon was, Apollo 15 Astronaut James Irwin once said *"God walking on earth is more important than man walking on the moon"*. But did God walk on earth? Is there even a God? With the advancement of science, have we moved past the need to believe in God?

"In the beginning, God created". Those first five words of the Bible raise the greatest questions for scientists, atheists, and people of all religions. They have filled lecture theatres, churches, and libraries. They have been examined under the microscope and searched for through the telescope. How you interpret those five words will determine your entire worldview.

It might surprise you that science now agrees with the first three words of the Bible *"In the beginning"*. That hasn't always been so. Those words raise the fundamental question of whether everything has always been, or whether there was a moment where nothing existed and then something existed. This is an important question because if there was a time when nothing existed, including time itself, then how did 'something' commence? The German philosopher Martin Heidegger phrased it this way *"Why are there beings at all instead of nothing?"* [1] And this is precisely where our conversation starts.

What or who caused something to come from nothing? It's difficult for our minds to imagine this because when we try to think of nothing, we are inclined to think of empty space, but empty space is not empty at all. It comprises what has been termed 'dark energy'. To imagine nothing is difficult to do, but that highlights the size of the problem.

———◆◆◆———

Prior to the 20th century, the predominant view was the universe was infinite. However, from about the 1960s, mainly through astronomy, a mountain of evidence has changed that theory. The universe is now considered finite and expanding, with the popular explanation being

4

that something the size of a subatomic particle exploded with such force and speed that our universe was formed, otherwise known as the Big Bang. While not everyone agrees there was a big bang, science and the Bible now agree on this one thing, that there was a beginning.

That conclusion is significant and naturally leads to a more important question, *'what or who caused the beginning of something?'* From everything we observe, and from basic common sense, we know that everything that began to exist had a beginning and a cause.

There are two main explanations to genuinely consider. Either the beginning of everything was due to a supreme being who existed outside the material universe, or the nothingness before time and space was its own cause.

The Bible's answer to what caused the universe is simple and explained in two words *"God created"*. While there is overwhelming agreement with the first three words in the Bible *"In the beginning"*, it's these next two words that cause debate. The Bible doesn't mince words. It claims there is a God and that God is the cause of the universe beginning from nothing. This assertion is incredibly important and if it's true, it has eternal implications. It shouldn't be accepted at face value, and it is right to ask for evidence and to question whether there are other valid naturalistic alternatives.

Some prominent atheist scientists have attempted to answer the question of something arising from nothing by suggesting that 'nothing' is actually something. Some atheists have theorised that because there is an equal amount of positive and negative energy in the universe, it adds up to zero and accordingly, we live in a perpetual cycle of nothingness which we choose to call something. Surely, we can do a bit better than that. A common criticism toward those who believe in God is they leave their brains at the church entrance and put them back in when they leave. Perhaps some do, but equally, we must encourage atheists to hold onto theirs as they worship at the altar of nothingness.

———◆●●————

The question of what or who started the physical universe can be pushed a lot further. How the universe started is one question, but what formed biological life and the invisible laws of mathematics and physics within the universe is a completely separate question and potentially far more complex.

It's one thing to ask how the fish tank came to be, but it's another thing to know how the water and the fish came to be in the tank, and how the water happened to be ideally suited to the fish.

How organic life came from inorganic chemicals is a profound question that currently has no valid explanation. There are theoretical answers, but no one can prove the theories. Dr James Tour is considered a world leading synthetic organic chemist, and is a Professor of Chemistry, Computer Science, Materials Science, and Nano Engineering at Rice University. Commenting on the question of how life might have formed from non-life, he notes *"Those who think scientists understand the issues of prebiotic chemistry are wholly misinformed. Nobody understands them. Maybe one day we will. But that day is far from today."* [2]

The question of how life commenced is similar to the question of how the universe commenced. There are only two options to consider. Life is either the result of a supreme intelligence outside the universe, or it is the result of random unguided processes. Which is the most logical conclusion, and which is best supported by the evidence?

If the Bible is correct, then it follows that scientific observation should support the existence of God as the most logical conclusion beyond reasonable doubt. That's an exciting prospect, because it means to a certain extent, the existence of God can be rationally tested, and that's an examination that should be embraced without fear of where it might lead.

————◆●◆————

I know this is a fairly heavy way to start a book but bear with me because we are asking some big questions.

Other than a headache, where does all this leave us so far? Well, it seems there is consensus on the first three words of the Bible

"In the beginning", but not on the next two *"God created"*. Three out of five words is 60% agreement, and that's not a bad start. If we can agree on those next two words, the rest of it is plain sailing. In fact, if you can logically agree with the first five words of the Bible, then the rest of the Bible naturally follows. However, the possibility that *"God created"* is certainly not something to be taken purely by faith. It needs to be assessed, and that is the challenge I invite you to embark on both in this book and in conducting your own research.

To an extent, this examination should be objective and without emotion. The evidence is what it is, and the conclusions should naturally follow. However, we also need to be aware there will be a point where we reach the end of our ability to understand or to search any further, and we must draw a conclusion based on what we can reasonably conclude. In that respect, we are required to exercise faith, whether that is the faith of the atheist who concludes there is no God, or the faith of those who conclude there is a God. Both exercise faith, and it's not a matter of science versus religion.

———◆●◆———

The Bible challenges us to examine the evidence for God, but it also says there is a point where faith is required, just as it is for the atheist. In the Bible, the book of Hebrews (chapter 11, verses 1-3) says *"Now, faith is the substance of things hoped for, the evidence of things not seen.... By faith we understand that the worlds were formed by the word of God, so that the things which are seen were not made of things which are visible"*. That's interesting. Even back then, people were pondering the question of how something came from nothing. The book of Hebrews challenged the readers to conclude the universe we see was made of something we cannot see, which is the very question we are grappling with. They were told the 'worlds' (the original word means the universe and time) were put together and ordered by God simply speaking them into existence and into order.

The readers of Hebrews weren't called to understand all the detail of how, but they were called to trust what God had said in the first five words of the Bible. We will never know everything. We will always be learning and always be seeking to understand. When the book of Hebrews was written, people had far less ability and tools to examine the universe, but based on what they could see, they were challenged to believe by faith that God formed the universe. So that begs the question, what is faith?

The first part of the passage in Hebrews states *"faith is the...evidence of things not seen"*. How can faith be evidence? The Bible's New Testament was first written in Greek and when we examine the original Greek words in this passage, we understand them better. The words mean *'that by which invisible things are proved and we are convinced of their reality'*. It can also mean *'an inner conviction based on proof'*. In other words, faith means we examine evidence to the point we can, and we then transition from weighing up information to forming a belief. Comparing this to a court examination, in a legal sense, we must look at the facts and then decide whether the Biblical account is true beyond any reasonable doubt.

At times, Christians are accused of having blind faith. I am sure that some do, but that's not what the Bible expects. It's a shame that 'faith' has become a term to describe those who believe without any evidence. That's not what it is, and the Bible never intended it to be applied in that way. The God who created the mind expects the mind to be used in exercising faith.

———•◆•———

How objective and open minded are we willing to be in examining the question of God? In a debate at Southampton University on 31 January 2019 between Professor John Lennox and atheist Peter Atkins, Atkins stated that even if he stood at the foot of the cross and saw with his own eyes Jesus die and rise from the grave, he would not believe in God, and would instead believe he was hallucinating. That's an astounding admission. It says that when it boils down, some people believe in things despite all evidence to the contrary. The Bible addresses this with an insightful and perhaps provocative statement. Psalm 14, verse 1 says *"The fool says in his heart, 'there is no God'"*. When evidence points to God, but we conclude there is no God, it's not an intellectual conclusion, rather it is a decision of the heart to refuse to believe. However, before we ask the question about our heart, we need to use our minds and examine the evidence.

We might at this point ask why does any of this matter? Is it worth investing time exploring the evidence for God? Is it really possible to be sure of the answer if we do? The author of *'War of the Worlds'* H.G. Wells answered the question in this way, *"If there is no god, nothing matters. If there is a god, nothing else matters"*. Pausing for a moment, let's think this through further. Imagine for a moment that instead of being the product of chance and accident, you were created by God and the true you is not only physical but spiritual. Imagine further that when your body gives up the ghost so to speak, you will meet the God who made you.

If there is a remote possibility that this could be true, then it should be the most important thing you could possibly think about or address in your life. Nothing else should matter. You should wake up thinking about it and go to sleep thinking about it until you reach a conclusion.

I recall some years ago, being in my downtown office in Wellington, New Zealand on the 6th floor when an earthquake happened. New Zealand is known for earthquakes, especially Wellington, but this one was decent. At the time, I was working on a legal issue which was important at the time, but when that quake started, everything else faded into insignificance. I found myself under the desk, staring at my brother who was standing under the doorway. As it went on, I began wondering how long the old building could hold together and my vision sharpened as I thought through what to do. No one else could make decisions for me at that time or help me. Once the shaking stopped, we quickly left the building and I grabbed someone in the hallway who was frozen with fear, and we hurriedly walked down the stairs. When we left the building, the street was full of people sitting on the road, trying to reach their families.

The point is that when my life was potentially on the line, nothing else mattered except getting to safety. The legal case I was working on which at the time was important suddenly became irrelevant and I don't even remember what it was. When it comes to our eternity, the moment we start to think about it should be like an earthquake. Nothing else should matter until we answer the question. It's too important to ignore. The potential ramifications are unlike anything else you will face in life.

Some might say yes, it's important, but there is no way we can ever really know, so what's the point in trying? However, respectfully that is a cop out and you're too intelligent for me to let you off the hook so easily. Every day, we form opinions and make decisions on things even in the absence of conclusive evidence. The lack of certainty does not prevent us from forming opinions or getting on with our lives. We take an umbrella with us because our phone tells us it might be a wet day, even though we know that many times the weather predictions are wrong. We get on a plane not knowing if it will crash. Without consciously thinking about it, we weigh up the evidence and conclude the plane will most likely arrive safely at its destination, although we can't be certain. We also conclude the pilot is appropriately qualified to be in the cockpit and didn't arrive to work with a hangover! None of this is certain, but it doesn't prevent us from drawing conclusions and making decisions.

If we are prepared to make decisions each day based on the limited information we have, then we should certainly be prepared to examine the evidence and form an opinion on the far greater question of our eternity. There may be a time and place for sitting on the fence, but this is not that time. It's just too important.

———◆●◆———

In my profession as a lawyer, I have conducted many investigations. At face value, the allegations against someone can appear compelling, but more often than not, as the investigation progresses, the evidence tells a different story.

A good investigator approaches issues objectively and unemotionally, not persuaded by peer pressure or the majority view. They allow the evidence to fall where it falls.

What about you? If the evidence for God is compelling beyond reasonable doubt, what will be your conclusion? Like the atheist Peter Atkins, have you already decided you do not want there to be a God, and therefore will not believe despite any evidence to the contrary? Or are you prepared to follow the evidence regardless of where it takes you?

As you read this book, what world view do you currently have? One world view tells us there is a God who made us. It's a view that explains who we are and our eternal purpose. The other world view tells us we came from nothing, for nothing, to nothing, and we are simply random accidents without meaning.

As we examine these two views, I encourage you to ask which one is best supported by the evidence. In doing so, make sure you do not succumb to peer pressure or the popular view. Stand back and look at the evidence as if you are doing so for the first time.

On this journey, we need to be clear about a couple of things. First, we need to set aside the notion that only religiously minded or weak-minded people believe in God. The camp of Atheism is full of brilliant people of all professional backgrounds, and the camp of God believers is also full of such people. Many amazing foundational discoveries in science, mathematics and other disciplines have been made by God believers. For them, believing in God was not unscientific. Instead, it gave them a starting point for understanding and researching their area of expertise.

The second thing to bear in mind is we cannot outsource our decision on the existence of God or on matters concerning our eternity. Saying we believe in God because our family does is not good enough. We are accountable for forming our own opinion on this issue.

As a young man at Law School, I was not prepared to accept the existence of God purely because someone told me or because my parents believed in God. I undertook an examination of the evidence, and I interrogated the Bible because it unashamedly claimed it was authored by God himself. Long story short, I got to the end of that journey and not only concluded there was a God, and the God of the Bible was real, but I personally encountered God and entered into a relationship with him. That was my journey, but it's one we each need to take. The Bible promises that each of us will discover and meet God if we undertake a genuine search with all our heart and mind. That's a Biblical promise I'd encourage you to test for yourself.

DIDN'T MY SCHOOLING PROVE EVOLUTION AND NULLIFY GOD?

"The first to present his case seems right, till another comes forward and questions him." (Proverbs 18:17)

Let's get a bit provocative to start things off. Let me start with an assertion which I don't expect you to take at face value, and I'd be disappointed if you did. The assertion is that most of what you know from school and university about macro-evolution (meaning one kind of creature evolving into another kind), and the origin of life is wrong. That's a serious assertion. If you just reacted to that assertion let me ask you one further question. Can you confidently prove macro-evolution, or do you believe it to be true because someone told you it was?

As an investigator, I like to deal with facts, so I don't raise these questions lightly. I raise them because having explored them, I have concluded that our education system has seriously let us down. We have left school with a warped view of the world, our existence, and our purpose.

Let's deal with an obvious hesitation or rebuttal you might have at this point. The rebuttal is you were taught in school that many 'missing links' have been discovered which demonstrate a transition from apes to humans, and accordingly, macro-evolution is a proven fact and there is no need to discuss it. I get it because that's what I was taught too. But was it true, or was it quite honestly, a load of nonsense?

Our schoolbooks were full of pictures of early human ancestors using tools, making fires, and chasing bison and mammoths. You might recall hearing about our evolutionary ancestors such as Neanderthal man, Nebraska man, Lucy, and Piltdown man. I recall those too. I remember first seeing those pictures and concluding it must be a done deal. There must be museums full of skeletons and partially developed ancestors. They must be digging up fossils that were part way through evolutionary stages by the truck loads all around the world.

Given the millions of evolutionary transitional stages between species that macro-evolution requires, I assumed there must be vast quantities of evidence in the fossil record. Archaeologists had done their homework and I merely needed to believe them. However, in later examining things myself, oh how terribly wrong I had been.

Without having to search far, I discovered many of the so-called 'missing links' comprised mere fragments of bones or partial skulls, and they have been found to be inconclusive, or in some cases have been dismissed as hoaxes.

For example, in the case of Neanderthal man, it was later concluded the skeleton was likely that of a man suffering from arthritis which explained his stooped over appearance. This is disturbing. As we will see later, even world respected palaeontologists have admitted there is no convincing evidence for the concept of macro-evolution and it worries them.

To help address this hesitation that comes from our schooling, let's briefly review some of the more common 'missing links'. In doing so, it's important to note we are not by any means suggesting all evolutionary scientists have dismissed these fossils. In some cases, they have used them to build an entire view of human ancestry and they hold strongly to those beliefs. However, the evidence is not at all conclusive, and is in some cases intentionally deceptive. In your education, you may have come across some or all of these examples.

Ramapithecus (allegedly the first man type ancestor)[3]

The conclusion now is that while it was *"once regarded as the ancestor of humans, it has now been realised (sic) that it is merely an extinct type of orangutan (an ape)."*

Australopithecus[4]

There is no consensus among experts on what the Australopithecus skulls are. Some consider they are skulls of an extinct ape. Others consider they are skulls of a species that was on its way to becoming human.

Lucy[5]

In the early 1980s a small part of a skull was found, and it was held up as the definitive 'missing link'. School books were full of this so-called convincing proof. However, this fossil has now been rejected. As Richard Leakey, a highly regarded fossil anthropologist notes, Lucy's skull is so incomplete that most of it is *"imagination made of plaster of paris"*.

In 1987, Dr Charles Oxnard, Professor of Anatomy and Human Biology at the University of Western Australia said *"Conclusions about Lucy being a missing link have been due to people with prior biased notions of what fossils would have looked like"*. The conclusion is that this partial incomplete skull was probably that of a common Pygmy Chimpanzee.

Homo Erectus[6]

Only parts of the Homo Erectus skull were found. While it is inconclusive what creature these parts belonged to, the shape and size was within the normal range of humans today and does not indicate it was a human ancestor.

Neanderthal Man[7]

We know this name well as it is used to describe our apparent ancestor who worked with fire and tools and had a thick skull. The original 'neanderthal skeleton' was stooped over and led some to believe it was an early human ancestor. However, subsequent analysis determined that the skeleton showed signs of severe arthritis which explained the stooped over appearance.

While many scientists continue to believe Neanderthal Man is an evolutionary human ancestor, the conclusion by many others is that these skeletons are not missing links but rather a 'cousin' of modern humans who became extinct.

Piltdown Man (Eoanthropus)[8]

The skull of Piltdown Man was publicized as the missing link for 40 years. In 1953 it was found to be an elaborate hoax. The skull comprised of a human skull cap with an orangutan's jaw deliberately glued on.

Nebraska Man[9]

While the impression is that an entire skeleton was found, Nebraska Man was actually based on one fossilised tooth, which was later found to be a tooth belonging to a pig.

Another argument in support of human evolution is the comparison of our DNA with that of chimpanzees. It would seem we have a lot in common, so does that not prove humans evolved from chimpanzees? Not in the slightest. As geneticist Steve Jones notes *"A chimp may share 98 percent of its DNA with ourselves but it is not 98 percent human: it is not human at all – it is a chimp. And does the fact that we have genes in common with a mouse, or a banana say anything about human nature? Some claim that genes will tell us what we really are. The idea is absurd."* [10]

This similarity of DNA across species actually points to evidence of a master designer, utilising a consistent software architecture across all life forms. This same rationale applies when we consider the range of technology we use every day. My computer, microwave and vacuum cleaner all have microchips which are built on a similar software architecture, but I don't conclude the similarity of microchips means my microwave evolved from my vacuum cleaner.

———◆●◆———

Where does our discussion take us so far? It leaves us less certain of the evidence for evolution we have perhaps considered as watertight up until this point.

Furthermore, there is a frustration that while much attention is given in the media and the classroom when a new 'missing link' is discovered, there is little attention given when it is later debunked. In fact, there is often silence. Many educational institutions even continue teaching some of these 'missing links' as part of the biology curriculum, including for example in New Zealand. [11]

The science community should be as prompt in correcting misidentified fossils as they are when announcing their discovery, but regrettably they are not and it damages the integrity that scientific discovery so desperately needs.

Given schools and universities exclusively teach the evolutionary worldview, students have not been encouraged to take a truly logical, evidence-based approach to the important question of our origin. As they are taught about evolution from an early age and are presented with pictures and alleged evidence of missing links and evolutionary fossils, there is no driving reason for them to ask the tough questions or challenge the underlying assertions as they move through the schooling system.

Naturally, with such education, students leave school, having never questioned the foundation of what they were taught, and they take that education into adulthood, adopting it as their worldview. Some go on to become teachers and scientists themselves, and still never question the foundation. How did we get to this? What is really driving it?

We need to approach this topic with fresh eyes and ask the obvious questions.

The importance of this struck me when I visited the American Museum of Natural History and looked at a large skeletal display of different sized horses. The line up of horses was presented as strong evidence for evolution. As I examined the display, it dawned on me that these were all horses! There was no evidence of a fish becoming a horse. They were merely different sizes of the same kind of animal. This was evidence of micro-evolution (meaning changes within a species), but not evidence of macro-evolution. In fairness, the museum may not have intended for the display to demonstrate macro-evolution, but it certainly used it to build the case.

The horse display was evidence of micro-evolution which means changes within a kind of species. We should not be at all surprised about that as it happens all the time – just look at yourself compared to your parents. But to extrapolate that out and claim a horse can turn into another kind of creature is nonsense. It hasn't happened in the past. It's not happening now, and there is no evidence for it. Some leading palaeontologists freely admit that is the case.

Now, I hear some objecting that this dismissal of macro-evolution is too simplistic, and I agree. It's too important to answer in such a brief summary, so we will examine this in more detail later on.

———————◆●◆———————

I fear our education system has failed generations of students in the way it has taught the origin of life and the universe. The classroom has forced a worldview that explains our existence as the product of chance, evolution, and randomness despite the absence of evidence. Students leave education believing it's a done deal, an open and shut case, and there's nothing to question. All the while, the truth is out there but the doors of our minds have been forced shut.

As a professional investigator I see this happen often. People I interview sometimes tell me a story to persuade me of their view. It often sounds convincing, but on closer examination and testing the underlying facts, I discover where there are holes and where there are lies. Only in testing and challenging things can we discover the truth, but we must be willing to accept that somewhere along the way we may have had the wool pulled over our eyes. No one wants to admit they fell for a scam, but in doing so, it's the first step to discovering the truth.

———————◆●◆———————

We live in an age that has been termed 'post-truth' and it's a dangerous age to live in. Post-truth is an approach to life that relies on emotion, personal beliefs and bias, instead of objective facts. Post-truth discourages debate and silences opinions that do not side with the cultural direction of our time, otherwise known as the 'cancel culture'.

In exploring the origins of life, scientists who dare to question the theory of evolution quickly feel the wrath of the post-truth community and the cancel culture.

How far we have fallen from rational discussion and exploration. As the Greek philosopher Aristotle is attributed to have said *"It is the mark of an educated mind to be able to entertain a thought without accepting it."*[12] We don't have to agree with each other, but we should be able to debate dispassionately, weighing up each other's views, and being prepared for others to question our strongly held beliefs. Another Greek philosopher Socrates summed it up when he said *"No greater misfortune could happen to anyone than that of developing a dislike for argument."* [13]

An unemotional analytical approach is critical when considering the topic of evolution and the existence of God. We need to be open to explore the very foundation of what we believe and objectively examine the evidence. It is insufficient to approach this with a post-truth perspective.

———◆◆◆———

At some point in our lives, regardless of our education and upbringing, we inevitably turn our mind to three important questions: where did I come from, why am I here, and what happens when I die? Our education system answers those questions by telling us we came from nothing, our purpose is to simply add to human achievement as best we can, and we have no existence beyond the grave. That's not particularly encouraging, and it leads to a sense of hopelessness. However, if the facts point to evolution and chance as the cause of our existence, then we have to be intellectually honest and accept those conclusions, regardless of how depressing that might appear.

When dwelling on life in the absence of God, King Solomon concluded with this statement in the book of Ecclesiastes (chapter 1, verse 2), *"Meaningless, meaningless. Everything is meaningless"*. He was wealthy enough to grant himself every pleasure his heart desired and he enjoyed the fame and adoration of all his citizens. There was no one wealthier or more famous than Solomon. From an evolutionary worldview, he achieved everything that could be achieved, and yet he looked on and declared it was all meaningless. He concluded that true purpose could only be found in knowing and living for his creator.

So where does all this discussion leave us so far in our journey? It has highlighted two important things. First, we need open minds when examining the claims about God and evolution. Second, we need to appreciate that this is not just an intellectual exercise. The conclusions we make speak not only to the existence or non-existence of God, but also to our purpose in life and potentially our eternity.

In the movie 'The Matrix', the whole world is asleep and plugged into a virtual world which they think is real. The main character Neo starts experiencing things he can't explain and starts questioning his reality. In one of the famous scenes, he meets people who have found the truth, and they give him a choice. The choice is to take a blue pill and go back to living the way he always had, putting those doubts behind him. The other choice is to take a red pill and discover the truth of who he really is. They warn him that in taking the red pill, he will know the truth and it's a decision he can never go back from.

In the movie, some people chose the blue pill because they didn't want their comfortable lifestyle to be disrupted. For them, it was easier to live in a dream even though they knew it wasn't real.

What about you? Like Neo, perhaps you have had some niggling questions about your existence or your eternity and have yet to be satisfied with the answers this world offers. As C.S Lewis said *"If I find in myself desires which nothing in this world can satisfy, the only logical explanation is that I was made for another world"* [14] So, which pill are you willing to swallow?

HASN'T SCIENCE KILLED
THE IDEA OF GOD?

"We Palaeontologists have said that the history of life supports [the story of gradual adaptive change] knowing all the while it does not" (Niles Eldredge)

When contemplating our existence, a common response is we no longer need to reference the idea of God as an explanation because science has worked it all out. While our ancestors clung to primitive religious beliefs, scientific discovery has enabled us to move on. If that's true, then its compelling, but how strong is that position?

In 1835 Charles Darwin visited the Galapagos Islands for five weeks as part of a five-year mission to survey the coast of South America. Darwin observed a range of species spread across various islands, and he noticed that certain characteristics within species were different, depending on the island they lived on.

This led him to form a theory that small changes have occurred across species for millions of years, and extrapolating back in time, he concluded that all life on earth came from a common ancestor.

In his book *'On the Origin of Species'* 1859, Darwin presented what he termed the 'tree of life' to graphically illustrate how he considered all species evolved from a common ancestor. You may recall seeing that illustration in school biology books.

People have queried whether Darwin believed in God and whether his scientific enquiries shifted his belief. While in his early years, Darwin described himself as a theist (meaning he believed in a creator God), he later confirmed he was an agnostic (meaning he believed he could not know whether or not God existed). In a letter dated 1879, Darwin wrote *"In my most extreme fluctuations I have never been an atheist in the sense of denying the existence of a God. — I think that generally (& more and more so as I grow older) but not always, that an agnostic would be the most correct description of my state of mind."* [15]

Darwin was not on a quest to disprove the existence of God, rather he was attempting to find a reasonable scientific explanation for the variety of lifeforms on earth.

Darwin did not tackle the question of how life came from non-life and that is an important distinction. His book was called *'On the Origin of Species'* and not *'On the Origin of Life'*. He had no clue about how life first arose and as we shall later see, nothing has changed. There is still no valid scientific explanation for how life started.

In summary, Darwin's theory was that as life progressed, vast numbers of mutations occurred over vast amounts of time and these resulted in vast numbers of species. He theorised there was an unseen hand he called 'natural selection' which guided the mutations through successive generations of species, enabling them to advance from one species to another. Darwin theorised that of the billions of mutations that might occur as species reproduce, the only ones that pass on to the next generation are those that help the species more readily survive in their particular environment.

Darwin made numerous observations on the Galapagos Islands

which he considered supported his theory. For example, he noticed there were variations in beak size between Finches depending on the Islands they were on and the food available for them to eat. On one Island, in order to reach the nectar, the Finches needed long beaks, so on that Island, Finches predominantly had long beaks. Whereas, on other Islands, they had short beaks. His conclusion was that this demonstrated adaptation of the species over time and proved his theory of evolution.

For his time, Darwin's effort in developing the evolutionary theory was impressive. He took a scientific approach by observing and then extrapolating his observations to develop an overall explanation. However, while he did well for his day, there were fundamental gaps in his understanding which were not his fault, but rather a consequence of the age he lived in. Darwin had no knowledge of the complexity of the biological cell and no knowledge of DNA, or the detailed software language embedded within it. These are critical elements we will return to as they seriously undermine the theory of macro-evolution. Although the theory continues to be taught as the dominant explanation for life, it is increasingly regarded as an implausible explanation.

---◆◆◆---

As we examine the topic of evolution, terminology is important. There are two types of evolution, and we need to be clear about which type we are discussing.

Micro-evolution describes the natural changes that happen within species over successive generations. When your parents welcomed you into the world, you were a combination of their DNA. While you reflect some of their characteristics, you are not a clone. The same will be true of your offspring.

This is an example of micro-evolution. It describes changes within a species. However, it is important to note that without exception, micro-evolution never leads to one type of species turning into another type. Such change is impossible because through successive generations of species, there is no new genetic information introduced. The only information that can possibly be passed on is the information necessary to make more of the same species.

The variation in Darwin's Finches fell into the category of micro-evolution. The Finches had the genetic information capable of producing either short or long beaks. To reach the nectar, the Finches needed long beaks, so only those with long beaks survived. This was micro-evolution at work, but note that the Finches remained Finches. There was no change from one species to another.

Darwin extrapolated what he observed with micro-evolution and theorised that given enough time, variations across generations could result in entirely new species, such as fish turning into birds. This is known as macro-evolution.

The idea with macro-evolution is that over vast amounts of time, with millions of very minor mutations across millions of generations, one kind of species can become another. A fish with gills can eventually become a bird with lungs.

Unlike micro-evolution, this has never been observed. The fossil record has been extensively searched across the planet and is still being combed for evidence, but so far is completely empty. That may come as a surprise to you. Despite the convincing drawings in school textbooks, there is no evidence in the fossil record to support macro-evolution and it worries some palaeontologists.

Furthermore, putting aside the fossil record, if macro-evolution is true and has been occurring for millions of years, we should see countless examples of transitional species living today, but we don't. Where are the millions of creatures alive today that awkwardly display their mutations as they transition from one species to another?

By definition, macro-evolution is continuing with every generation of every species, and there should be vast numbers of transitional forms observable around us, but there isn't. Not one. This presents a significant problem for Darwin's theory. When the evidence isn't under the ground, and it's not above ground, where exactly is it? Something is not adding up.

What we see in the fossil record is very simple. It is full of all kinds of fully formed species, with no examples of one species evolving into another.

We are certainly not short of a decent sample size to make our findings either. In fact, we are inundated with fossils all over the world. While writing this book, I walked on a beach in New Zealand and stumbled upon a fossilized fish embedded in a rock about 8 feet off the ground. Fossils are discovered everywhere, and yet no one has ever found a fossilized creature on its transitional journey from one species to another. Again, when the evidence isn't under the ground, and it's not above ground, where exactly is it?

———•••———

Darwin knew he would receive backlash, but as a professional scientist, he was not deterred. In pursuit of truth, he was willing to raise questions despite the criticism he might receive.

Darwin was not the only scientist willing to pursue truth in the face of academic and religious opposition. As an example, we owe a lot to Galileo Galilei who experienced backlash when he introduced his theory about the sun and the planets. In 1610 he published a book theorising that the earth and planets revolve around the sun. With his new powerful telescope, he had more to support his theory than Darwin had because unlike Darwin, he could observe and then test his theory by personal observation.

The Catholic church did not take kindly to this theory as it disrupted their belief that the earth was the centre of the universe. Even with Galileo later explaining how the ocean tides supported his theory, the Catholic church would not listen and he was placed under arrest. Galileo's books were banned, and he was not permitted to teach his theories. Perhaps of most concern was that Galileo offered on numerous occasions for philosophers to look through his telescope, but they refused to do so. In writing to a friend, Galileo noted *"...these philosophers shut their eyes to the light of truth"* [16]

Unfortunately, it appears we have come full circle and have closed our eyes to re-examining our strongly held beliefs. Scientists who question macro-evolution quickly experience the wrath of the academic community. Some scientists have claimed that research funding is only available for those who follow the rules. The message to them is clear, don't rock the boat.

I suspect Darwin would be disappointed to learn how scientists are persecuted in this way for bravely challenging the status quo. He was one of those scientists and it's because of his bravery that we have the theory of macro-evolution in the first place!

—————◆◆◆—————

Darwin took his profession seriously. He knew the value of developing hypotheses and the importance of being prepared to discard them if the evidence proved them wrong.

Darwin was careful not to overpromise. He realised that such a significant theory would need to be supported by evidence and that such evidence should be in abundance within the fossil record. Darwin conceded that the biggest failure of his theory would be revealed if no intermediate variations between species was found in the fossil record.

In his view, the number of intermediate forms in the fossil record should be enormous. In his book *'On the Origin of Species',* he noted *"(Since) innumerable transitional forms must have existed, why do we not find them imbedded in countless numbers in the crust of the earth?"* (page 162) and *"Why is not every geological formation and every stratum full of such intermediate links? Geology assuredly does not reveal any such finely graduated organic chain: and this perhaps is the most obvious and gravest objection which can be urged against my theory. The explanation lies, as I believe, in the extreme imperfection of the geological record"* (page 293)

On the one hand, Darwin acknowledged the fossil record should be full of transitional forms, but because it isn't, he was content to say the geological record is not perfect. With all due respect to Darwin, that is a major cop out and demonstrated personal bias to his theory at the cost of objective analysis. Unfortunately, it seems the effort he gave in testing his theory did not match the rigor he had in developing it.

We can all agree with Darwin that the geological record is imperfect but given the millions if not billions of mutations and transitional stages of evolution that supposedly occurred, one would expect to find at least a few intermediary fossils, even a handful would be helpful. In his book *'Beyond Natural Selection',* Robert Wesson notes, *"Large evolutionary innovations are not well understood. None has ever been observed, and we have no idea whether any may be in progress. There is no good fossil record."* [17]

Macro-evolution makes some very bold claims about what happened in the past. There is no evidence of it occurring today, so the only information scientists can rely on to test the theory is the fossil record. If there was no fossil record at all, then scientists could arguably say there is no evidence that macro-evolution did not happen, and it would potentially lend more weight to their theory. However, perhaps regrettably for the theory of macro-evolution, there is an enormous sample size in the fossil record, large enough to confidently test the theory, and it is there for all to see. Evolutionary biologists are confronted with evidence in the fossil record that tells a very different story to the theory they hold to.

———————◆•◆———————

In addition to the lack of evidence, the theory of macro-evolution runs counter to a well proven scientific principle called the principle of uniformity. We rely on this principle every day without ever pausing to think about it. The principle of uniformity says we can make solid assumptions about tomorrow based on our experience of today, and we can equally draw sound conclusions about the past based also on our experience of today. Occasionally, significant events will occur that disrupt this principle, such as natural disasters, but they are exceptions to the general rule.

We can safely conclude there will be a sunrise tomorrow because there was one today and there have been sunrises since records began. Satellites are placed in orbit because there is an assumption that gravity and the earth's rotation will continue today as it was yesterday.

Based on this principle, we can also conclude that tomorrow, humans will give birth to humans, dogs will produce dogs, and fish will produce fish. From our experience there is no reason to believe this will not happen. Equally, based on the principle of uniformity, we can make a very strong assumption that prior to today, reproduction occurred as it does today.

When considering reproduction of species, there is nothing to indicate the principle of uniformity was broken yesterday, last year, 1000 years ago or a million years ago.

Given the uniformity of reproduction we see today with every species on earth, the burden of proving the principle of uniformity was broken in the past would require an exceptional amount of evidence because it would go against everything that science so heavily relies on and everything we observe.

However, there is no such evidence. Furthermore, we are faced with a huge amount of evidence in the fossil record that supports the principle of uniformity and by simple reasoning disproves, or at the very least undermines the idea of macro-evolution.

———•••———

One of the confronting realities in the fossil record for evolutionists is something called the 'Cambrian layer'. The Cambrian layer is a layer of sediment within the earth where vast numbers of fully formed species suddenly emerge with no prior evolutionary stages in earlier layers within the earth.

From an evolutionary worldview, it makes no sense and presents an enormous challenge. There should be millions if not billions of fossils demonstrating their transitional stages between species prior to the Cambrian layer, but there are none. The Cambrian layer seems to suggest that a large range of species simply appeared on earth at a point in time with no prior evolutionary stages occurring. It almost seems to support the Bible's account that God made all creatures 'according to their kind'. Now, that's an interesting theory to examine.

Commenting on whether the fossil record contains transitional stages of evolution, previous Palaeontologist of the Natural History Museum, Colin Patterson, noted *"I will lay it on the line – there is not one such fossil for which one could make a watertight argument"* [18] So much for the school text books and the colourful illustrations.

If I was seeking to explain the absence of evidence in the fossil record, I would perhaps present an argument as follows. We know that fossils only form when there is an instantaneous event comprising of vast amounts of water and debris. Creatures are quickly buried with such pressure that all oxygen is removed and there is no opportunity for decay. Accordingly, there must have been catastrophic events at different times in history that caused only those generations alive at the time to be fossilised. Therefore, we would not expect to see prior generations of creatures in the fossil record because they all died and decayed. However, there are two problems with this. First, whenever such catastrophic events occurred, the numerous transitional forms of species alive at the time would have been fossilised, but we see no such fossils. Second, to embrace the idea of a catastrophic event lends serious weight to the Bible's claim that there was a global flood, and that is not a door any evolutionist wants to open. Incidentally, the global flood described in the Bible would have provided perfect conditions for fossils to form and would be consistent with the evidence we see in the fossil record.

———— •◆• ————

Given all this background, it seems our education system has built its foundation on incredibly shaky ground, void of evidence and layered with problems, and yet it teaches students the debate has been settled. It hasn't, and the lack of supporting evidence is deeply concerning. At best, the education system is naïve or mistaken. At worst, it is pursuing an agenda. These are not accusations from non-scientists or God promoters. They come from evolutionists. Niles Eldredge of the American Museum of Natural History noted that *"We Palaeontologists have said that the history of life supports [the story of gradual adaptive change] knowing all the while it does not"* [19]

During a public lecture presented at New York City's American Museum of Natural History on 5 November 1981, previous Palaeontologist of the Natural History Museum, Colin Patterson said *"... I've tried putting a simple question to various people and groups of people: 'Can you tell me anything you know about evolution, any one thing that you think is true?' I tried that question on the geology staff in the Field Museum of Natural History, and the only answer I got was silence. I tried it on the members of the Evolutionary Morphology Seminar in the University of Chicago ... and all I got there was silence for a long time, and then eventually one person said: 'Yes, I do know one thing. It ought not to be taught in high school.'"*

These are serious objections to macro-evolution by people who have devoted themselves to the theory, and accordingly their objections should be taken seriously. However, when approaching this topic, there is a frustration that doesn't need to be there but is there none the less.

Due to the education system being steeped in evolutionary theory, students of science are grounded in a belief that Darwin was right and there is vast evidence to support his theory. For them it is a confirmed worldview which does not need to be questioned because there is nothing to question. This is a difficult thing to unravel with people who are so certain of what they have been taught.

Consequently, scientists who question Darwinism are seen as crackpots, irrational, or driving religious agendas and are categorised in the same way as moon landing deniers or flat earthers.

However, in recent years, the number of scientists who doubt and challenge Darwinism is growing and many of these scientists are not religious. They have looked at the evidence dispassionately and do not consider it supports Darwinism. With the advancement of scientific knowledge, they consider the probability of macro-evolution occurring is no longer only improbable, it is impossible. In challenging these bedrocks of evolutionary belief, such scientists face at times emotional and at times fanatical opposition.

Acknowledging this difficulty, David Gelernter (Professor of Computer Science at Yale School of Engineering and Applied Science) noted *"Darwinism is no longer just a scientific theory but the basis of a worldview, and an emergency religion for the many troubled souls who need one".* [20] Gelernter also notes that when evolution is challenged, there is often an emotional and angry response because it is a challenge to peoples' religion. [21]

When people display an emotional or dismissive response to serious questions or challenges to their beliefs, it highlights that such people are not in search of truth. They have already made up their minds and will only entertain ideas that support their worldview. This is a danger that everyone has irrespective of their worldview. When we form opinions on subjects, it is natural to only listen to or seek out information that supports our opinion. This is known as 'confirmation bias'. However, if our search is genuinely for truth, we must seek out and follow the evidence wherever it leads, and must avoid dismissing challenges to our beliefs simply because we do not want to accept we may be wrong.

———————•••———————

It's time for a serious examination that puts aside the sacred cows and casts aside prejudice, emotion and personal beliefs. Only with such an approach will we arrive at a logical, well-thought through conclusion. If evolution is true, fine, lets deal with it. But if it's not, then let's deal with that too.

If you got this far in the book and are prepared to examine your worldview with objective, rationale eyes, then I take my hat off to you.

With that mindset I now encourage you to examine some of the evidence in the next few chapters and see where it takes you on your journey for truth.

Having personally examined the evidence, my view is there is a God, and he is responsible for the intelligence and complexity of life we see in our world. In addition, I have also concluded God has personally revealed himself to mankind in ways which we can objectively and subjectively test.

Here's a personal challenge. If having examined the evidence, you conclude beyond reasonable doubt there is no God, then so be it. You can move on and explore alternative explanations for who you are and why you exist. But if you conclude God is real, then it's a game changer. It then raises significant questions such as why you are here, can you know God, what does he want of you, and what happens after you die? I can't think of a more important and exciting investigation. Remember, a good investigator follows the evidence regardless of where it leads, so let's continue the investigation with an open mind.

WHAT EVIDENCE WOULD WE ACCEPT FOR GOD'S EXISTENCE?

"The more I study science, the more I believe in God" [22]
(Albert Einstein)

Before we continue looking at evidence for the existence of God, let's start with the question of what type of evidence we would accept to prove his existence. It is important to first define the criteria that might indicate there is a God, otherwise, you may never arrive at a point of conclusion.

Perhaps you have never turned your mind to this question, or perhaps you believe God exists because the Bible or a religious book says so, or because you have been told so. But what makes you so sure? What is the basis for your belief?

If God exists, is it reasonable to expect he would provide us with observable evidence of his existence? The idea that an all-powerful God who created the universe would reveal anything of himself to people on a tiny globe in the Milky Way galaxy is an amazing thought. If it is true, it places a sense of importance not only on this planet but on humanity.

If there is such a God who needs nothing and was before all things, what interest would he have in reaching out to those he created? These are important questions because they touch on the idea that if we conclude there is a God, we might also be able to discover something about his nature or character, and our purpose. That would shift the conversation from an intellectual one of God's existence, to a relational one of whether we can actually know God personally.

We are jumping ahead of ourselves, but this is important because there is no point in arriving at a conclusion that God exists if it has no practical application for us personally.

Introducing the concept of evidence when discussing God, and in particular the God of the Bible, may surprise you. Perhaps you have assumed those who follow the Bible act in faith and that faith is the result of wishful thinking or is the opposite to reason and science.

However, the Bible paints a very different picture. It expects us and challenges us to examine the evidence. It makes the case that the evidence for God is so strong that the onus is on us to prove there is no God as opposed to proving there is.

In the Bible's New Testament, the book of Romans (chapter 1 verse 20), states *"For since the creation of the world God's invisible qualities— his eternal power and divine nature—have been clearly seen, being understood from what has been made, so that people are without excuse".* The Bible not only says there is a God, but it also claims the evidence for his existence is abundantly clear, so much so that when we die, no one will have an excuse for not having believed He exists.

If you are stopped by a police officer for speeding, despite passing 5 signs that all say 'roadworks slow down', it is no excuse to say you did not see the signs. The Bible says it's the same when it comes to the evidence around us. We can choose to shut our eyes to the obvious, but the truth will catch up with us. To be aware of the signs and choose to ignore them is something in law we call 'wilful blindness'. In other words, we choose to shut our eyes to the obvious.

Whatever our belief system, we all have the same road signs. The Bible says the signs point to a creator, whereas our education system says they point to nothingness.

Sometimes we can be staring at the same thing but miss the most obvious truth. It is like the account of Sherlock Holmes and Dr Watson who go camping together. In the night, Sherlock wakes Watson and asks him to look up at the sky and explain what he sees. Dr Watson gives a detailed account of the galaxies and the billions of planets. He then asks Sherlock what he observes and Sherlock pauses and says *"someone has stolen our tent"* [23] It's the same with the question of how we came to exist. We all have the same facts to consider. Whether it's the complexity of life, the fossil record, the intricate language of DNA, the countless galaxies, or the fine tuning of our planet to support life, like Sherlock asked, when you consider these things, what do you see? Sometimes, it's easy to miss the thing right in front of us.

If God exists and if he wanted his existence to be known, what would he do? What would it look like? In considering this question, we must be careful. The danger is we set our own rules for what we expect God should do and then choose not to believe in him when our self-defined conditions are not met.

I have heard people say they cannot believe in a God who permits all the suffering in the world. For them, the existence of suffering means there is no God. I understand the emotion behind that statement, but it is illogical. It's like saying because people die in car crashes, I can't believe in the people who invented cars. Suffering exists, but it does not logically follow to say that God does not. To draw that conclusion means you have already defined who God would be if he did exist. What if God was mean and nasty and wanted people to suffer? The fact we may not like that has no relevance to whether he exists.

What people are really meaning is they don't understand why there is suffering, and that if there is an all-powerful God, why would he permit it. That's a fair enough question.

However, if people conclude the presence of evil or suffering in the world proves there is no God, they must then grapple with what to conclude when they consider the presence of good or happiness. If the presence of evil suggests there is no God, then what does the presence of good suggest?

In examining the evidence for God, we must avoid falling into arguments that are based on emotion or personal preference, even where that emotion stems from a genuine hurtful experience. Equally, we must also set aside our pre-determined concepts of God and open our eyes to things perhaps we have not considered before.

———————————•••———————————

I have heard some say that if God showed himself visibly to them, then that would be sufficient evidence to prove his existence. However, if this was the test for God's existence, he would need to reveal himself visibly to every person at every time in history. As that has not happened, we could therefore choose to conclude God does not exist. However, that would involve drawing a significant conclusion based on a self-defined test. A more objective conclusion would be to acknowledge our test has not been satisfied, and then ask what other type of evidence would be sufficient to prove God has revealed himself.

Would it be sufficient evidence if God visibly revealed himself to a large number of people at one time in history and those people documented their eyewitness encounter? Would it add further weight if those people then willingly died for what they said they had seen and heard?

You have probably guessed what I am alluding to. In the Bible, the New Testament is full of eyewitness accounts of the life, death and resurrection of a man called Jesus Christ. That matters for many reasons, the most important being that the Bible states Jesus is God who came to earth to live among those he created. If that's right, it's a big deal. Under extreme pressure and torture, all of his disciples held to that belief, and all except for the apostle John, were killed for doing so.

When the New Testament was written, it not only mentioned many eyewitnesses by name, but it also referred to an event where over 500 people saw Jesus at the same time after he rose from the dead. The writer of the book of 1 Corinthians also mentioned that many of those people were still alive at the time of writing. In other words, the facts could be verified by asking the eyewitnesses who were still alive.

At this point, you could say you were not there so you can't conclude whether such events actually happened. However, if our assessment of truth and history is based on whether we personally experienced it, then we have very little basis to believe anything from the past. We believe Hitler committed suicide in a bunker at the end of the second world war, but none of us were there to witness it, and we rely on the testimony of others. We believe Isaac Newton first thought about the law of gravity when he observed an apple fall from a tree, but none of us were there to verify it. There is no photographic evidence of my great, great, great grandparents and I did not meet them personally, but I have no doubt they existed. I never met William Shakespeare, but I have full confidence he existed and that he wrote the likes of Hamlet and Romeo and Juliet.

An investigator doesn't need to have personally witnessed an event in order to form a reasoned conclusion on whether it happened. In fact, an investigator is usually chosen because they were not an eyewitness and are considered independent. Good investigators form reasonable conclusions due to the weight and reliability of the evidence they are presented with and not due to their own personal involvement. It's how our legal system works. It's how historians operate. We rely on it.

Let's return to the question again. What evidence would you be willing to accept for the existence of God? It's surprising the things we are happy to be confident of with very little evidence.

If you arrived on a desert Island and found intelligible patterns on the rocks, you would likely conclude someone had been there before you. As Professor John Lennox of Oxford University noted *"A few small marks on a flint are enough to tell an archaeologist that he is dealing with an artefact, and not just a piece of weathered stone. Inferences to intelligent agency are made as a matter of routine in disciplines such as archaeology, cryptography, computer science and forensic medicine"* [24] At the most basic level, when we see things that appear designed and ordered, we automatically conclude there is a designer. It doesn't take much at all. However, when we look at the most complex things in our universe, such as organic life and the language of DNA, we are encouraged to do the exact opposite and believe it all came about by chance. This demonstrates a disturbing inconsistency of logic.

Belief in God without evidence is not an attitude the Bible expects us to have. To the contrary, the Bible challenges us to refute the volume of evidence it says exists all around us.

The Bible presents two main categories of evidence that it asserts prove the existence of God. The first is something scholars have called 'General Revelation', or for our purposes we might call it 'General Evidence'. This refers to what we see, hear and feel of the natural world around us. Examples are DNA, organic life, fossils, the cycle of the seasons, and the food, oxygen and water which is all ideally suited for us to exist. It's called 'General' evidence because it doesn't require any special encounter with God. Instead, these facts are available for everyone to observe. The Bible's perspective is we are confronted every day with vast quantities of General Evidence. It asserts the evidence is so overwhelming that on this evidence alone, God will hold us to account if we choose to reject the idea that he exists.

The second type of evidence the Bible alludes to has been called 'Special Revelation' or we might call it 'Special Evidence'. This refers to the Bible's claim that God has personally made himself known to his creation. While some might dismiss this as subjective experience, we should not be so quick to do so. The Bible itself claims to be a message from God, and that claim can be robustly examined.

———◆●◆———

There is an account in the Bible in which the Apostle Paul spoke to people about the topic of General Evidence. In the book of Acts (chapter 14), we read the account of Paul performing a miracle and then the crowd wanting to offer sacrifices to him, thinking he was a god. At that time, the Greeks worshipped many different gods such as Neptune the god of the sea, and Zeus the god of thunder. Paul rebuked the crowd and told them not to worship him. He explained it was time to put aside myths and instead to use their intellect and look at the General Evidence for God's existence. In doing so, he drew upon evidence they could all relate to, namely how their day-to-day existence was totally dependent on God's provision. Paul said:

> "...We are bringing you good news, telling you to turn from
> these worthless things to the living God, who made heaven

and earth and the sea and everything in them. In the past, he let all nations go their own way. Yet he has not left himself without testimony. He has shown kindness by giving you rain from heaven and crops in their seasons; he provides you with plenty of food and fills your hearts with joy." (Acts chapter 14, verses 15-17)

We all have the same General Evidence. Some conclude it is all the product of time and chance. However, Paul argued the existence of the seasons and the food that was ideally suited to our physical requirements are all evidence for God's existence.

According to the Bible, the universe contains a vast amount of evidence. Psalm 19, verses 1-4 states *"The heavens declare the glory of God; the skies proclaim the work of his hands. Day after day they pour forth speech; night after night they display knowledge. There is no speech or language where their voice is not heard. Their voice goes out into all the earth, their words to the ends of the world."*

Psalm 19 is not intended to be read as colourful poetry. It presents things in a way that requires the reader to either accept them or reject them. It declares there is a God, he is the creator, and the heavens are evidence of both. According to this passage, people are not required to have a scientific background or special equipment to make a conclusion of God's existence. From day 1, people could look up to the heavens and reasonably conclude there is an all-powerful God.

On a clear night, people throughout the world can look up and witness the same heavenly objects the Psalmist wrote about. It doesn't matter what language we speak or which continent we live on. When we look up and see the vast array of the heavens, we observe the glory and creative power of God.

The heavens are not just full of inanimate objects. By their very existence, they 'proclaim', 'speak', and 'display knowledge', and in that sense they have a 'voice'. Although atheists would say they are simply random lights in the heavens, the Bible says that by their very existence they are an enduring declaration of God's incredible creative power.

Down through the years there have been mystic religions which have lured people in on the basis that truth and spiritual things are hidden and will slowly be revealed the deeper they enter the religion. This philosophy is known as the occult which means *'kept secret', 'beyond the range of ordinary knowledge'.*[25] In stark contrast to this, Psalm 19 says God's existence and power is not hidden. It is displayed before our very eyes. As a result, while we might point our finger to the heavens and challenge God to prove his existence, his finger points back, challenging us to deny his.

————◆●◆————

In the classroom we are taught we came from nothing, by nothing, for nothing and to nothing. In contrast, the Bible declares the universe and our existence came from the mind and power of God. It was purposeful which means we also have a purpose.

The Bible describes God's creative power this way in the book of Genesis (chapter 1, verse 3) "And God <u>said</u>, 'Let there be light'..." According to the Bible, our reality commenced because God in his absolute power spoke it into being.

It's fascinating the Bible describes the first creative act as one of language, *"And God <u>said</u>"*. With advancement of science, we now understand that the very make up of who we are is encoded through the language of DNA. Who would have known it would take until the late 20th century for science to discover something the Bible had alluded to 6,000 years ago?

However, it doesn't end there. The book of John in the Bible reveals that this act of creative power was more personal than we could ever imagine. John (chapter 1, verses 1-3) says *"In the beginning was the Word, and the Word was with God, and the Word was God. He was with God in the beginning. Through him all things were made; without him nothing was made that has been made."* And in verse 14, *"The Word became flesh and made his dwelling among us. We have seen his glory, the glory of the one and only Son, who came from the Father, full of grace and truth".*

If you didn't notice, that's a mic drop moment. We earlier read the first verse in the Bible *"In the beginning God created the heavens and the earth."* But now, we discover the one who spoke the universe into existence with unfathomable power was not an impersonal god or the product of randomness, rather it was a person who entered the pages of history as a man, Jesus Christ, who's other title is *"The Word".*

Introducing a personal God into the equation shifts the discussion of our existence from how we exist to why we exist. If it's true that the God who made us became one of us and lived among us, then that screams of purpose and relationship.

———————◆•◆———————

So far, we have been talking about the General Evidence of God's existence based on the things we can observe in nature and the universe.

The other part of General Evidence relates to things we can't see, and more specifically, it addresses the uniqueness of humanity compared to any other creature.

The Bible says that unlike all other creatures, God made humans in his image. This means that while our biological makeup is amazing evidence of God's existence, there are additional characteristics unique to humans that we can't put under a microscope, but they exist, nonetheless. These characteristics are evidence of God's involvement in our existence as opposed to the product of randomness.

It goes without saying that as humans, we stand apart from all other creatures. That doesn't mean we are the strongest or fastest creature. There are many things other creatures can do that we can't. We overcome these limitations by using our imagination and intellect at a level that only humans have. We can't fly so we build planes. We can't swim to the depths, so we build submarines. We can't outrun horses, so we drive cars. All of this is impressive and is on another scale to the creativity and intelligence of any other creature. However, as amazing as this is, there are other aspects unique to human beings that speak to the fingerprint of God in a more profound way.

The fact you are reading this book and asking yourself questions about your own existence is a remarkable example of General Evidence for God's existence. Humans are unlike any other creature in that we have the mental ability to observe and investigate our existence and then draw conclusions on whether or not there is a God. As far as we know, we are the only creature who has this ability [26]. We take this for granted and seldom stop to consider how remarkable this is.

The ability to think about the fact that we are thinking is a profound ability. Known as 'metacognition', this unique consciousness is so profound and so beyond any other creature, that it is evidence of a supreme designer with intelligence far beyond our comprehension.

It is this ability to think about the fact we are thinking that caused the philosopher Rene Descartes to conclude philosophically that he existed. In contemplating that he could think about thinking, he declared *"Cogito, ergo sum"*, translated *as "I think, therefore I am"* [27]

Even in 2024, science readily admits that no one understands consciousness. [28] Anil Seth, a cognitive and computational neuroscientist, and co-director of the Sackler Centre for Consciousness Science at the University of Sussex in Brighton, UK notes *"It's still just fundamentally mysterious how consciousness happens"* [29]

To grasp how profound the unique consciousness is that we have, and then conclude it is the product of nothingness, randomness and chance is quite astonishing. It would seem that such a conclusion is one of the heart and not the mind.

————◆◆————

ARGUMENTS FOR GOD'S EXISTENCE

"Men occasionally stumble over the truth, but most of them pick themselves up and hurry off as if nothing had happened"
(Winston Churchill)

We have come a long way in our understanding of the universe. Our telescopes help us look further up, and our microscopes further down. However, with the increase in knowledge, we are left with the same question of how it all came to be. This question is beyond the brightest minds, greatest microscopes, and profoundest philosophers.

As the complexity of our universe becomes more apparent, the belief in randomness and chance as explanations for what we see appears less and less plausible. While some have argued that science has buried God, the contrary argument is science increasingly reveals Him.

Putting aside microscopes and telescopes, there are several arguments for God's existence which science has no convincing explanation for. Some of the well-known ones are summarised below.

The fact the universe exists at all
(known as the cosmological argument)

When my children were young, I enjoyed tucking them in at night, and sometimes I would ask them what the toughest question was they could think of. They did a fairly good job with questions such as *'who made God?'*, and *'what happens when we die?'* One question that particularly impressed me was *'what was there before God made everything?'*

That's a good question, and it can be phrased in this way – *'why is there something rather than nothing?'* The fact there is something instead of nothing is a profound mystery for philosophers and one which science does not adequately explain. If the Big Bang did occur, it does not answer the question of what was before the Big Bang. The idea that nothing existed and then exploded is difficult to visualise, even for those with the greatest imaginations or for those we tuck in at night.

To believe that something arose from nothing requires much more than a leap of faith. Even Stephen Hawking noted *"It is difficult to discuss the beginning of the universe without mentioning the concept of God..."* [30] Allan Sandage who discovered quasars noted *"...God to me is a mystery but is the explanation for the miracle of existence – why there is something rather than nothing"* [31]

Experience tells us everything that commenced to exist had a cause. It follows that because the universe began to exist at some point, it also had a cause. If the Big Bang happened, there first had to be something to bang, so we are left with the same problem of what caused that something to exist. All scientific theories reach a dead end at this point.

Regarding the theory of the Big Bang, I'm not a scientist but humour me for a moment. The theory is that something the size of a subatomic particle exploded with such force and speed that our universe was formed. Think about that. Something so small that we can't see, contained the mass of all the planets and stars in the universe. Apparently, we were also part of that subatomic particle.

When we go on family holidays, it's my job to pack the car, and when the kids were young that was a challenging task with toys and pushchairs and suitcases. Things would keep coming out to be packed that I didn't even know we owned! My rule was I didn't start packing until everything was by the car ready to be packed. However, that rule didn't often work. Inevitably, after some brilliant packing on my part, an awkwardly shaped box would find its way out to the car, and I'd have to start over. The point is, there's only so much squeezing and condensing that can occur.

I don't know about you, but I have immense difficulty believing all matter in the entire universe can be packed into something which is not only smaller than a regular car or even a matchbox car, rather it's smaller than an atom. Admittedly, it's only a theory, but it certainly tests my imagination. In any event, it doesn't answer how the subatomic particle came to exist in the first place.

In 2021, NASA launched the James Webb Space Telescope into space. This highly advanced equipment has been taking images deep into the universe and some of those images are now challenging the Big Bang theory. As the telescope focussed as far into the universe as possible, the prediction was that galaxies would be smaller and eventually there would be nothing. However, the images reveal massive galaxies even bigger than the Milky Way, and the findings are causing astronomers to rethink theories of cosmology [32]. What this means is the theory of the Big Bang is certainly not settled, and other theories and models will continue to emerge.

So, what was the first cause that led to the existence of the universe? Atheists argue that introducing God as the answer is contradictory because God would also need a cause to exist. However, that demonstrates a misunderstanding of God, particularly the God of the Bible. There certainly are examples of people believing in gods that were themselves created, such as Apollos or Thor, but those gods had a beginning. The God of the Bible is entirely different. The first book of the Bible states simply *"In the beginning God"*. In other words, God is outside creation. He had no beginning and therefore had no cause. Everything that exists came from him. He is the first cause that had no cause himself.

The intelligibility of the universe

Albert Einstein is quoted as saying *"The most incomprehensible thing about the universe is that it is comprehensible"*, or words to that effect[33].

At University I enjoyed taking maths classes, but as the formulas became more complex, I couldn't imagine a practical situation where I would ever use them, so I shifted to other topics. I must confess those formulas found their way out of my brain fairly quickly.

After I left University, I discovered that those abstract formulas we wrote down on paper are the foundation for much of the design we see in nature. Think about that. Completely unrelated parts of the natural world comply with the same abstract mathematical formulas. If randomness was our god, that should not be the case.

Professor John Lennox, Professor of Mathematics at Oxford University (emeritus), summed up this profound mystery as follows *"...the intelligibility of the universe is grounded in the nature of the ultimate rationality of God: both the real world and mathematics are traceable to the mind of God who created both the universe and the human mind. It is, therefore not surprising when the mathematical theories spun by human minds created in the image of God's mind, find ready application in a universe whose architect was that same creative mind"* [34]

In the Bible, the book of Hebrews (chapter 1 verse 3) says of God, *"He sustains all things by his powerful word"*. In other words, God not only created the laws of mathematics and physics, but he also sustains those laws so our universe can continue to exist. It is the constancy of those laws that enables us to make important scientific discoveries. Sir Isaac Newton didn't invent the law of gravity, he discovered it, and we can now accurately predict lunar eclipses based on Newton's rule of gravitational attraction. Newton could take the scribblings on his maths sheet and make accurate predictions, not because the universe was random, but because he expected it to follow the rules.

There are laws the universe complies with, and we can write those laws on a piece of paper. That presents a picture of intelligence and design and is completely opposite to the idea of a random chaotic universe.

The intricate mathematical design in nature points to a brilliant and creative God who designed an architectural framework for all creation which rests on mathematical consistency.

From experience, we know that logic and order come from intelligence and design, not from randomness. Given there is intelligibility in our universe, it is most logical to conclude there is an intelligence behind it. A very strong argument for God indeed.

The rationality of our minds

When I was at University, I made some extra money selling laptops. Bear in mind, at this time, you couldn't walk into a store and buy one. It was 1991, and the laptops were basically bricks, computers with a lid. The battery lasted about 2 hours, the screens weren't colour, and the memory was so small it relied on the use of floppy disks to provide the software. With all those limitations, people loved them. They were amazing for their time.

When I powered up those laptops and conducted a demonstration, I had full confidence they would do what they were supposed to do because some very bright minds had designed them. If I had been told those laptops were put together by young children connecting random wires, I certainly would not have showed up to give demonstrations to potential purchasers!

Computers have come a long way, and we rely on them every day. Whether it's our cars, our laptops, or our smart watches and phones, we use them, never questioning whether they are reliable. We do so, because we know they were designed to perform a task, and we trust the competence of the people who designed them.

As amazing as computers are, they pale into comparison with the complexity and power of the human brain, and yet we are taught at school that our brains are the product of total unguided random processes with no design involved. Oh dear. How can we conclude a designer was involved when considering a simple computer, and then conclude randomness was responsible when considering our brains which are infinitely more complex.

As you read this book, you take for granted that your brain recognises the words and letters and translates them into meaning that you can understand. Even more astounding is that while doing so, you can ponder the deep mysteries of the universe and of your own existence. This is an extraordinary ability. The conclusion is obvious - you can think with structure and intelligence because there was an intelligence that structured your brain. If your brain came about due to billions of mistakes and random mutations over millions of years, with no order and design, you have no reason to rely on it. It's simple. Chaos creates chaos and nothing else.

Complexity points to purpose (also known as the argument of teleology – 'telos' meaning purpose)

My wife is a potter, and potters have all kinds of small tools that help them shape the clay. When a courier package arrives at the door with some small tools inside, I have at times picked them up and wondered what their purpose is. How will they be used? Someone designed those little tools with a specific purpose in mind. As simple as those tools are, they have a level of complexity to them with their wooden handle and sharp metal end. We do not look at simple tools and conclude they came about through randomness. That would be stupidity.

How does this relate to God's existence? The point is, if something appears to have a purpose, the most logical conclusion is it was purposefully designed, and therefore points to a creator.

It only takes the slightest hint of purpose, and we conclude there was an intelligence involved. Whether it's a pen, a paperclip, or a matchstick, we know they have a purpose and so we conclude there was a designer. It's logical. It makes sense.

Although we know this to be true, our education let us down. In the classroom, we learnt about DNA and how its purpose is to provide software for the formation of our cells. We were also taught that all our body parts, such as our heart and lungs, have a specific purpose and without them we cannot survive. And yet, while we know a matchstick has a purpose and therefore was created, we were taught to ignore the apparent purpose and design in our bodies and conclude instead that we are the product of total randomness with no designer involved. Quite frankly, that is an astounding failure of logic and its high time we see it for what it is. Complexity points to purpose and purpose points to a designer. It always does, no exceptions.

Our sense of what is right and wrong

In April 1945, Supreme Commander of the Allied forces in Europe, General Dwight Eisenhower visited Ohrdruf, the first Nazi concentration camp to be liberated. What he saw horrified him. United States Army General George Patton arrived before Eisenhower and was so sickened by what he saw that he threw up. [35] Patton forced the mayor of the town and his wife to walk through the camp and see what had been happening while they had turned a blind eye. The following day, both the mayor and his wife were found hanging in their home from an apparent suicide, and a note nearby read *"We didn't know. But we knew."* [36]

When Eisenhower visited Ohrdruf, he sent a cable message to General George Marshall, the head of Joint Chiefs of Staff in Washington and said *"The things I saw beggar description. ... The visual evidence and the verbal testimony of starvation, cruelty and bestiality were so overpowering as to leave me a bit sick ... . I made the visit deliberately, in order to be in a position to give first-hand evidence of these things if ever, in the future, there develops a tendency to charge these allegations merely to 'propaganda.'"* [37]

On April 15, Eisenhower wrote to his wife and said *"The other day I visited a German internment camp. I never dreamed that such cruelty, bestiality, & savagery could really exist in this world! It was horrible"* [38]

Eisenhower was so overcome by the moral atrocities, he sent an urgent message to General Marshall, telling him to come to the camp and bring with him journalists and members of congress. He knew the importance of documenting the atrocities for future generations in case the world forgot how evil the heart of man can be.

Forgive me for being blunt, but if there is no God, none of these things were wrong, nor were they evil. If there is no God, and we are the result of random processes, then the concepts of good and evil are meaningless. There is no basis whatsoever to argue our actions are evil. Good and evil do not exist because there is no higher reference point to make them so. We have no ultimate accountability, and no moral obligation to one another for our actions.

From an evolutionary perspective, Hitler was simply asserting his belief that the German Aryan race was the supreme race, and those he considered inferior should be eradicated. As abhorrent as that viewpoint sounds, it is the logical conclusion of evolution. Evolution encourages destruction of lesser species and survival of the fittest.

The point is, we know the cruelty of the Nazis was not only wrong, but it was also evil, and our hearts and minds affirm that to be true. Eisenhower was shocked by what he saw because evil does exist, and we know it when we see it.

Because we know what evil is and what good is, the question is how do we know and how do we all, for the most part, agree with each other? Where do we get a shared sense of right and wrong, good and evil?

Some might argue we all agree with certain moral standards because it is mutually beneficial. However, even though we might agree on mutual moral standards, we often don't act according to those standards because we seek out our own benefit and not the benefit of others. We all know that lying is wrong and that telling the truth is mutually beneficial, but we engage in lying because we consider it is to our advantage. People know adultery is wrong but engage in it because they consider the personal benefit outweighs the moral obligation to their spouse.

The reality is, that imprinted within us is a set of moral laws that we know to be true, and we endeavour to keep those laws as best we can. We don't need to teach our children these laws because they are printed on their souls. That is not to say they do not break these laws. Indeed, as parents, we need to regularly correct their actions. However, whether we are children or adults, we all have an internal compass that continues pointing north, whether or not we choose to follow.

We also feel a sense of guilt when we act contrary to these laws because we know intrinsically that in that moment we are a law breaker. No one needs to tell us. We know because our conscience is our judge, and our conscience holds up the moral law as a mirror to our soul.

Chance does not create morality, nor does it explain feelings of guilt or remorse for breaking moral laws. The Bible tells us in the book of Romans (chapter 2, verses 14-15) that God has imprinted his law on our hearts so that we intuitively know what is right and wrong. We feel guilty when we break those laws because we know the laws are true. Just as God designed humanity with the fingerprint of DNA, he equally designed us with the stamp of morality and gave us an inner conscience to assess our actions against that law.

IS EARTH A LITTLE
TOO CONVENIENT?

"I am convinced that God does not throw dice"
(Albert Einstein)

How are you finding our journey so far? Maybe you get the arguments and see some merit in them, but you want more facts and figures. I get that, so let's jump in.

You may not know it, but you are a very fast reader. How is that you ask? As you read this book, you are travelling through the vastness of space at 107,000 kilometres per hour as the earth orbits the sun. A very fast reader indeed!

It's one thing to talk about the marvel of the universe and our galaxy, but let's consider our own backyard for a while. Let's talk about earth. In the vastness of space, and in the midst of 100 billion stars in our galaxy, we comfortably live on our planet, enjoying its warmth, atmosphere and all the conditions we so desperately need to make life possible.

Perhaps you have never paused to consider how remarkable this is. The probability that following the Big Bang, earth randomly formed with all the necessary parameters for life to have a fighting chance, is incredibly small. Bear in mind, we aren't talking about the probability of life forming. That is far more complex and requires a separate conversation. We are only considering the probability that earth formed in a manner that made life remotely possible. The probability of this happening is astronomical, and it is difficult to explain in a way that sufficiently presents the size of the problem. So, let's give this some context.

———◆●●———

In World War I, a bullet fired from a German rifle headed straight toward the heart of 17-year-old Leonard Knight. Leonard walked away without a scratch because the bullet penetrated a pocket Bible he was carrying. It stopped 50 pages from the end. What are the chances of that happening? It seems extremely lucky, but it's not really.

Taking a conservative calculation, the size of the pocket Bible compared to the front side of Leonard's body meant there was possibly a 1 in 200 chance the stray bullet would hit the Bible instead of any other part of the body. That's still a lucky escape, but those odds aren't ridiculous. To put it in perspective, the chance of winning the New Zealand lottery through purchasing one ticket is approximately 1 in 380,000. That's starting to up the game of probability.

Some years ago, I personally experienced something so improbable that it's hard to believe. New Zealand gets some wild weather on the west coast and on such a day, I joined my brother at the beach to engage in some long line fishing. My brother was kind enough to lend me his fishing rod and tackle. If you don't know much about fishing, just imagine a fishing rod with a line on it and at the end of the line, attached to a tiny round circle is a bunch of hooks and lures to attract the fish.

Joining others who were at the water's edge, I cast the rod far out into the sea, and shortly after, the line became stuck. After pulling a few times, it came free and I wound it back in but to my dismay, the tackle and the hooks had gone. They had snapped off the line.

I walked up the beach and my brother kindly gave me another set of tackle which I attached to the line. Now bear in mind, I did not keep track of where I had cast the line the first time. Returning to the water's edge, I cast the line out again. Within a minute I noticed I had caught something and quickly pulled in hard, fighting the waves. As the line reeled in, I was shocked to discover I had pulled up the very tackle I had lost! What are the chances I thought to myself. But it didn't end there. On closer inspection, I discovered the fishing hook on the new tackle had gone through the tiny circle that was on the end of the old tackle. It was a perfect catch. In that vast sea, with the waves and the wind, the hook had gone through a tiny circle less than 5 millimetres in diameter. I was astounded.

I would not know how to calculate the probability of that occurring. One in a million? Maybe one in ten million? Who knows. But here's the confronting fact. Even those odds are nothing compared to the probability of earth forming randomly to contain all the ingredients for a life sustaining planet.

The bullet hitting the Bible and the fishing hook catching the tackle were amazing, but they were one off events. The difference when considering the likelihood of earth forming, is that it needed incredibly unlikely events to all happen one after the other in perfect sequence. If one necessary event did not happen, then it was game over. If you did statistics at school, you might remember that probability gets exponentially harder when one event relies on another occurring.

With that background, let's consider some numbers. Astrophysicist Erik Zackrisson from Uppsala University in Sweden conducted computer modelling of the numerous requirements for our planet to exist with all the right ingredients for life. His calculations were based on the assumption that the Big Bang occurred. He concluded the probability of this happening is 1 in 700 quintillion[39], or 1 in 700,000,000,000,000,000,000. That's a big number. What this means, is if there were 700 quintillion planets, only one of those planets could potentially be suitable for sustaining life, and that just happens to be ours. We need to stress again that this is only the probability of a planet forming that could possibly sustain life. The probability of life emerging from non-life on such a planet is a far more complex calculation.

In mathematics, there is a point at which probability is no longer regarded as worth mentioning as it is essentially a fairy tale up there with the Easter Bunny and Santa Claus. That number is approximately 10^{-50}. Erik's number isn't quite that improbable, but it's not far off.

In September 1971, 84 of the world's top scientists met at the Byurakan Astrophysical Observatory, Yerevan, Armenia, for a conference on Communication with Extraterrestrial Intelligence (CETI), jointly sponsored by the U.S National Academy of Sciences and Akademiia nauk SSSR, the Academy of Science of the USSR. They tried to calculate whether it was possible that extraterrestrial life could exist elsewhere in the universe. Their conclusion was that life on earth is a unique event and the probability of life existing anywhere else is so small that it is a "miracle rather than a statistic". They also concluded the existence of life is the result of information as opposed to chance or unaided energy. In other words, an outside force was responsible for the formation of life.

There is certainly a lot more thinking going on in the academic community about our origin than we are led to believe. Why this type of research and data is not presented in our schooling is a mystery, and reinforces the suspicion that truth is not the end game in evolutionary science.

———◆◆———

We take for granted that the planet we live on only exists because of numerous finely tuned parameters. If any one of these were slightly different, we wouldn't be here. We don't tend to think about this a lot because we just get on with living. But let's take a moment and do just that.

Below is a sample of many criteria that had to have precisely occurred when earth was forming so that life could potentially arise.[40] The technical term for this is the 'anthropic principle'.

Like me, you may not understand all the underlying scientific language or references in these examples, but you'll get the idea.

When reviewing these conditions, bear in mind we don't get to pick and choose which one was most necessary. They all had to happen one upon the other.

Existence of Carbon

Carbon is critical for life to exist on earth. It's a non-negotiable requirement. Your body is approximately 20% carbon. While carbon is critical, the conditions had to be near perfect for it to form. Mathematician and Astronomer Sir Fred Hoyle noted that the 'nuclear ground state energy levels' had to be fine-tuned with respect to each other and that if they varied by 1%, it would not work, and carbon would not have existed. Although he was an atheist, this one discovery alone led him to state that it looked like *"a super intellect has monkeyed with physics as well as with chemistry and biology"* [41]

Existence of Stars

In an age where our cities are lit up at night, it's difficult to truly grasp the magnificence of the heavens. Perhaps you have been camping or had a chance to be away from the city lights and marvelled at the billions of stars that shine across the heavens. The Milky Way is so full of stars that it gives the impression of a bottle of milk having spilled across the heavens, which explains its name.

Many civilisations have gazed up wondering what those lights in the night sky are. With our telescopes and spacecrafts, we now know a lot more about them, and have the privilege of seeing beautiful images of distant galaxies. When we gaze at the heavens, most of us probably do not comprehend how perfect the conditions had to be for the stars to form in the first place.

For stars to form, the ratio of the 'nuclear strong force' to the 'electromagnetic force' had to be perfect. Theoretical Physicist Paul Davies notes that if the ratio had differed by 1 part in 10^{16} then no stars would have formed at all. That's a huge number that we can't really get our head around. One author explained that the likelihood of stars forming is like a marksman with one shot, aiming and hitting a target at the other side of the universe. [42] The conditions truly had to be near perfect. That's either an extraordinary stroke of luck or it was deliberately designed.

Distance from the Earth to the Sun

Some people are summer people and others are winter people. Personally, I'm a summer guy. There's nothing more refreshing than going for a run along the beach and diving into the lake to cool down or having a picnic on the boat and finally succumbing to the heat with a jump overboard. Where I live in Taupo, New Zealand, the summer temperatures don't climb too much more than 30 degrees Celsius, and the winter temperatures bottom out at about -2 degrees. Both extremes are fairly comfortable.

Other places in the world have much greater extremes, and people endure temperatures of -20 or even lower, while hotter climates can reach over 40.

The extraordinary thing is the temperature across the planet is within the range that life can survive in. Yes, there are extremes, but they are exceptions, and for the most part, with appropriate clothing, we can survive. Our oceans are not boiling, and our trees are not spontaneously catching on fire. Life carries on and we never stop to consider how amazing this is.

The heat we desperately need relies entirely on the sun, a ball of gas 1.3 million times the size of earth and 148 million kilometres away. The sun constantly sends it's light and heat toward earth. It had to be there for life to start on earth, and it must remain there in exactly the right spot for life to continue.

148 million kilometres is a long way. You might consider, that give or take 2 million kilometres would not make a big difference on earth, but that's not the case. If the sun was 2% further away from earth, it would be devastating. Earth would not receive enough heat, and the planet would freeze. Life would not be possible. Equally, if the sun was 2% closer to earth, the glaciers would melt, the oceans would cover the planet, and the heat would kill all life. Now that's something to reflect on next time you sunbathe.

Rotation of Earth

As I go about my day, I don't often pause to think how the planet I live on is rushing through space, orbiting the sun at a speed of 107,000 kilometres per hour, and is rotating on its axis at approximately 1,600 kilometres per hour. Sometimes I stand up and feel lightheaded, but I've never thought 'my goodness, I wish earth would stop rotating so fast!'.

The speed of earth's rotation is incredibly important. It turns out that 1,600 kilometres per hour is an important speed to make life on earth comfortable. In fact, if the earth rotated slightly faster, it would cause drastic events.

If the speed increased, eventually gravity would not be able to counterbalance the spin of the earth and we would have a hard time staying grounded. Our normal windy days would become catastrophic events.

If the earth spun merely 1 mile per hour faster, the water level around the equator would rise a few inches. [43] As it increased speed, eventually entire countries would be under water.

At the other extreme, a decreased rotation would make for longer days and nights. During the day, the surface of the earth would heat up too much, and then cool down too long at night, causing plant life to struggle for survival.

Somehow, earth rotates at a speed that is just right for a habitable planet. The 'coincidental' necessities for a life sustaining planet are certainly mounting up.

Nature and Supply of Water

Apparently, the unguided, unintelligent and random process of evolution conveniently resulted in a range of animals and plant life that all relied completely on water, and water just happens to be the most bountiful thing we have on earth. I say this tongue in cheek, but the point remains. The properties of water and the vast supply of it is absolutely essential for life.

We drink it, sweat it, and bathe in it. It cleanses the sky and rains down on the plants. It's constant cycle keeps the ecological system working. It can only do all these things because its composition allows it to be a liquid, a solid and a gas, all at the right temperatures.

Our bodies comprise approximately two thirds water, and while we can survive for some time without food, we will only survive a few days without water. It moderates our temperature and is the perfect substance to carry essential nutrients and chemicals throughout our body. Water enables our bodies to discard waste and toxins through sweet and urine and ensures our blood can flow through our veins. All our bodily systems depend on water.

According to evolution, no one told a blob of random molecules to form organs and bodily systems that depend on water. Evolution didn't even know what water was. And yet here we are.

Water has a wide temperature range from when it freezes to when it boils, and this is crucial. We wouldn't want water freezing or boiling when circulating through our veins.

Water also has an amazing feature in that it freezes from the top down. We can skate on top of a frozen lake and yet underneath, the fish survive.

The ability for water to evaporate is not only interesting, its critical for the planet's ecological system. Evaporation ensures the plentiful supply of salt water is filtered and dispersed throughout the world.

We all know that on the periodic table, the name for water is H2O. Two molecules of hydrogen and one of oxygen. If the properties of hydrogen and oxygen were slightly different, biological life would not be possible. [44]

That's a few things to think about next time you're hosing the garden and wiping the sweat off your brow.

Magnetic Shield

When watching Star Trek as a kid, it was always exciting when the captain ordered 'shields up'. An unseen forcefield was activated and protected the ship from enemy torpedoes. Amazingly, earth also has a natural protective shield around it, and it is critical to the planet's survival. It extends from deep within earth's inner core, and out into space.

The sun emits incredibly dangerous particles, solar winds and radiation which would destroy all life if there was no protective shield. The authors of 'The Privileged Planet' sum it up as follows "Like a *giant magnet, earth's geology conspires to create a magnetic 'shield', which protects earth's atmosphere from solar wind particles and low-energy cosmic ray particles"* [45] Our own forcefield. How cool is that!

Earth's Moon

On 1 April 2002, NASA released a statement claiming they had upgraded the Hubble telescope and could confirm the moon was indeed made of cheese. It made for a humorous April Fools.

When the Apollo 12 crew left the moon, they intentionally crashed the lunar module into the moon as an experiment, and this triggered what was known as a 'moonquake'. The seismometers recorded vibrations much longer and larger than had been anticipated which indicated the moon was less dense than was expected. A conspiracy quickly grew that the moon was hollow, and now a simple google search reveals the conspiracy has grown to some now believing the moon hides within it a spaceship.

Conspiracy theories aside, there are some amazing facts about the moon that demonstrate how important it is for life on earth.

The moon is 384,000 kilometres from earth and about 27% the size of earth. It is responsible for the ocean tides, and if the moon was much larger or closer to earth, those tides would become a serious problem.

If the moon was closer or larger, it would slow earth's rotation, causing much longer days and nights and extreme temperature fluctuations on the surface.

If the distance between earth and the moon was halved, we would get an impressive view of the moon's surface, but the downside would be ocean tides approximately 8 times higher than we are used to. It would also result in increased gravitational pull and would start to break up earth's crust, resulting in large earthquakes and volcanic eruptions.

On the other hand, if the moon moved away from earth, we would experience decreased ocean tides and earth would slowly stop spinning.

A quick look at the moon's surface with all the meteor holes shows we live in a rough neighbourhood and our moon has borne the brunt of those celestial travellers. If the moon was further away, we would face these intruders alone.

The authors of *'The Privileged Planet'* summed it up this way, *"The relationship between earth and its moon is so intimate that it's probably best not to think of earth as a lone planet, but as the habitable member of the earth-moon system. This partnership not only makes our existence possible, it also provides us with scientific knowledge we might otherwise lack"* [46]

Earth is a perfect observatory

The spacecraft Voyager 2 was launched on 20 August 1977 from Cape Canaveral Air Force Station in Florida, and 12 years later it reached Neptune, the outer planet of our solar system, having travelled 4.5 billion kilometres. That's a long way, but in comparison to the size and length of our Milky Way galaxy, its merely a brief bus ride.

On a clear night, the vastness of our galaxy can be seen, but it's hard to comprehend how large it is. Along with planets, gas, and dust, NASA estimates our galaxy comprises approximately 100 billion stars. Travelling at the speed of light, it would take approximately 100,000 years to travel across the galaxy. [47]

Despite its size, we get to cast our eyes across the galaxy with a front row seat. How is it that our view is not drowned out by the light from all those stars? It turns out earth is perfectly placed in the galaxy to be a natural observatory of not only our own galaxy, but also the wider universe. The authors of *'The Privileged Planet'* note that if there was slightly more starlight in our view, we could not observe the universe as we do.

Earth is also perfectly positioned in our own solar system to enable important scientific discoveries to be made, particularly during solar eclipses. Solar eclipses only happen because the distances between the moon, sun and earth are just right. A little more or a little less and this would not be possible. What's more, earth's atmosphere is just right so it acts as a crystal-clear lens which we look through as we gaze upon the night sky.

It's as if someone built a perfect observatory and placed it in the perfect position in the galaxy. However, while that is impressive, it would be pointless if the earth was not also the home of life-forms with sufficient intelligence to make the most of this natural observatory. And what do you know? It just so happens to be the case.

What does all this tell us?

Earth is too good to be true. We shouldn't be here. It doesn't make sense. Against all probability, our planet exists and it's in the perfect position, with the perfect qualities and perfect environment for a range of life forms to exist. Earth's existence is a miracle of immense proportions.

To their credit, some atheistic scientists have admitted that earth is essentially a miracle. However, instead of following the evidence and concluding it points to a creator, an alternative explanation has been put forward. The idea is that perhaps our universe is merely one of millions of universes, and our one just happens to be the one where the conditions for life exist. This is called the 'multiverse theory'. It sounds interesting, but there are three fundamental problems with this idea.

The first problem is there is no evidence to prove there are other universes, nor can the theory be tested, so it remains merely an idea. The second problem is that even if there were multiverses, it does not remove the question of God's existence, as God could have chosen to make multiverses and to make one where life could exist. The third problem is that it still fails to address the question of what caused the universe in the first place. The multiverse theory is nothing more than an avoidance from acknowledging the compelling evidence for a creator.

The multiverse theory makes for a good Marvel movie, but that's really where it belongs. There's too much at stake to come up with fanciful theories. We need to deal with what we actually know.

The real question is whether looking at our universe, it is more logical to conclude there are billions of universes all formed by accident, or whether the universe we live in is one formed by an intricately clever and intentional creator. [48] When considering the unlikelihood of a life sustaining planet emerging from nothing, the most logical and reasonable conclusion is it demonstrates the existence of God.

COULD LIFE ARISE
WITHOUT GOD?

"Everyone's clueless on this, and no one wants to admit it"
(Dr James Tour) [49]

Let's assume for a moment that against all probability, the conditions necessary for life on earth came about by chance. With a perfect environment and all the ingredients in place, how difficult would it be for life to then commence by accident?

In our school science class, we were told that creating life from non-life was not a big deal. Millions of years ago, somewhere on earth, there was a primordial pond comprising of water, hydrogen, methane, ammonia, and carbon dioxide. In one lucky strike, a bolt of volcanic lightning hit the pond and abracadabra, amino acids were formed which then assembled into proteins. Then, fortunately, this soup stayed exactly where it was for a few million years, and what do you know, a living cell was formed and then replicated itself. Easy.

We were also told that in 1953, a scientist mimicked the conditions of primordial soup in a laboratory and was able to create the building blocks of life. However, while it's true, his experiment did form several amino acids, no proteins were formed. To suggest his results had any comparison to life is like saying I mimicked the writings of Shakespeare by typing a comma on the keyboard. Sure, it's a start, but seriously?

Admittedly, some time ago there was a lack of understanding about the composition of a living cell. No one knew that a 'simple cell' contained complex DNA and intricate machines that unwind the long DNA strand, read it, and copy it.

Our textbooks say that with no intelligent intervention, primordial soup formed amino acids, then proteins, and then the membrane of a cell. It then wrote its own software, and then built its own microscopic machines to read and replicate the software to form new cells. Umm, did anyone just notice a pig fly past the window?

Even if a protein could have formed in the primordial soup, there is a monumental gap between the formation of a protein and a living cell. Science class left out that part of the story. The gap is enormous. It's like asking how the Sistine Chapel in Rome was formed and being told that all you need is a chisel and a paintbrush, and then in the words of Seinfeld, 'yada yada' you have a chapel.

It was not until recent years I discovered how many scientists now consider the soup theory to be nonsense. That doesn't mean they believe in God, rather they just don't buy the theory anymore, similar to many scientists who have asked for a refund on the evolutionary theory.

———◆◆◆———

The idea that life came from non-life is called 'abiogenesis'. Dr James Tour is a synthetic organic chemist, and is a Professor of Chemistry, Computer Science, Materials Science, and Nano Engineering at Rice University in Houston. According to Dr Tour, no one on earth has any idea how life started. That's quite a statement coming from an expert in the field.

From everything we know to be true, only life produces life. Rocks and minerals do not turn into life. Despite trying to replicate it, brilliant scientists with sophisticated laboratories in perfect conditions, have never made it happen. It simply doesn't work. Life does not come from non-life, period.

That being said, let's humour ourselves for a moment. If theoretically, life could arise from non-organic material, what is the probability that it would actually happen?

Calculating the probability of a living cell forming from inorganic material is far too difficult to calculate or explain. So, let's step back from the question of a living cell and ask a much simpler question. The question is, what is the likelihood that the first building block of a cell would arise by chance?

For a living cell to form, it first needed fully functioning proteins. Surely that should be simple enough right? How difficult could it be for proteins to form by chance? Unfortunately for evolutionary theory, it turns out that the formation of merely one protein is in reality mathematically impossible.

In his book *'Signature in the Cell: DNA and the evidence for intelligent design'* [50] Steven Meyer explains that the probability of just one functional protein (comprising of 150 amino acids) arising by chance is 1 in 10^{164}. To put that in perspective, Steven explains that if one amino acid chain formed every second with all the right ingredients present, it would take longer than the assumed age of the universe to form one protein. If you're not astounded by that, you need to read it again.

Let's be generous. Let's assume that against all odds, a protein was formed. The problem is that one protein means nothing. It doesn't get you anywhere toward life or a living system. The formation of that protein would need to be successfully repeated over and over, time after time to form enough proteins. If a mistake happened along that sequence, or one protein didn't beat the odds of forming, the life experiment would have been over, and the process would have to start over from scratch. Each protein had to form first time around, without mistakes. Are you getting this? Remember, this is just the problem of forming proteins, and in the journey to a cell, this is supposedly the easy part.

But that's not all. The bigger problem is that even if all the proteins formed, we are still nowhere close to a living cell. At that point there is no genetic information, no protective wall for a cell, and no microscopic machines to copy the genetic information. If we only have a few proteins and are hopeful for life to arise, it's like saying I found a jar of screws, so I basically have a Lamborghini ready to go.

Other scientists have conducted similar calculations to Steven Meyer's. Astronomers Fred Hoyle and Chandra Wickramasinghe in their book *'Evolution From Space'* (Simon and Schuster, 1981) calculated the chance of getting all enzymes needed for a cell at 1 in $10^{40,000}$ and as they noted *"an outrageously small probability that could not be faced even if the whole universe consisted of organic soup"* [51]

In *'Undeniable: How Biology Confirms Our Intuition That Life Is Designed'*, [52] Douglas Axe comments that when these types of

63

probabilities are discussed, it is not appropriate to explain them as highly improbable, rather they are impossible.

If life did start without God, no one knows how it happened. Dr Tour summed this up and noted *"Those who think scientists understand the issues of prebiotic chemistry are wholly misinformed. Nobody understands them. Maybe one day we will. But that day is far from today."* [53]

Dr Tour went on to speak of the complexity of the information within human DNA and noted *"The code is analogous to the difference between the Library of Congress and a big box of alphabetic letters— the library has a huge amount of embedded information while the random box of letters has little. So origin of first life is the 'nail holding the coffin closed' on the emergence of biological evolution. Without that first life, or simple cell, which requires the four molecule types plus information, all proposals regarding biological evolution are without the base of life. And it is difficult to discuss biology without life"*

Dr Tour explained the difficulty of creating one living system and proposed a scenario where *"The world's best synthetic chemists, biochemists, and evolutionary biologists have combined forces to form a team—a dream team in two quite distinct senses of the word. Money is no object. They have at their disposal the most advanced analytical facilities, the complete scientific literature, synthetic and natural coupling agents, and all the reagents their hearts might desire. Carbohydrates, lipids, amino acids, and nucleic acids are stored in their laboratories in a state of 100% enantiomeric purity"*. In his view, even given a few billion years, they would not succeed in creating life [54]

With this confronting information, allow me to be a little cheeky. To borrow a boxing analogy, at this point, after successive blows, chance and evolution are no longer on the ropes, they are out for the count or at best, on the mat moving in and out of consciousness. However, to use evolution's mantra, this is survival of the fittest, or fight to the death, so let's keep going and finish the job with a few other considerations.

To build a house, we need detailed architectural plans. To develop a computer game, we need detailed software coding. To make a movie, we need a script. To make a cake, we need a recipe. Everything we see in our world that has a purpose and appearance of design, was brought into existence by an intelligent being who developed a concept, and then brought together the necessary resources to make it a reality. It's what we know to be true. We rely on it.

However, while we know this is true, when it comes to the astounding variety of life on earth, evolutionary theory tells us there was no designer, no blueprint, no purpose, and no intelligence.

We are told to believe this, and yet, we now know that our DNA reveals unequalled design, software coding, intelligence, and purpose. It alone is beyond anything mankind has managed to create with all its intelligence and resources. The complexity of DNA makes our best efforts look like child's play.

Such complexity does not arise without intelligence, so where did it come from? You can't rationally attribute it to chaos, chance, and long periods of time. To do so goes against everything you and I know to be true.

While that is the case, let's continue examining the theory, that given enough time and mutations, complexity can arise from chaos without any intelligent intervention.

———————◆◆◆———————

Natural selection kills great ideas

The story of macro evolution relies on the concept of 'Natural Selection'. In essence, Natural Selection means that during reproduction, if an unhelpful mutation occurs, the creature will die, but if it is a helpful mutation then the creature will survive. That helpful mutation will then pass on to the next generation and maybe in a thousand years, another perfectly suitable but accidental mutation will add to that one, and so on and so on.

There are countless problems with this idea. A fundamental problem is that anywhere along this generational timeline, as soon as one of those successive mutations is unhelpful or not perfectly improving the previous mutation, then Natural Selection says the creature dies, and its back to square one. It doesn't go back one step in the process. It must start from scratch. All the supposed advancements that occurred are dead in the water. In short, the mistakes can't have their own mistakes. The mistakes must be perfect, each and every time.

However, we really are being too kind to this idea of Natural Selection. The greater problem is that mutations are by definition, not helpful. They make a creature weaker, not stronger. If a fish mutates to losing part of its gill on the way to developing a lung, it is weaker, and Natural Selection says the weak die and the strong survive.

Although this alone is an insurmountable problem for Natural Selection, there is a far greater problem, and it concerns the topic of information. For a creature to change on the outside, there must be a change on the inside to the creature's DNA, and this requires a careful re-writing of the software code.

A creature is limited by the DNA of its parents. That's the way it works. Without a major addition of genetic information from an external source, there is no possibility whatsoever for changing the creature's DNA software.

It is impossible to change a gill into a lung without a massive change to the software encoded in the creature's DNA, and such change does not come about by mutation. In fact, mutation is evidence of a loss of information or corrupted software. It is certainly not evidence of an enhancement.

If you know the smallest amount about computer software, you will know that if you remove a small part of it, the programme does not get better, it becomes corrupted. Equally, to make improvements to the software, there needs to be an addition of information from an external intelligent source. That's why we receive regular software updates on our computers. There are brilliant minds behind those updates, adding new information.

We have been talking conceptually, but let's bring all this together into a tangible example. Let's explore the very thing you are using to read this book, the human eye. Was it designed by a brilliant architect, or did it slowly evolve over millions of years from chaos and random mutations?

In 1966, a debate occurred between prominent mathematicians and biologists at the Wistar Institute in Philadelphia. Mathematician Stanley Ulam noted that based on his calculations, it was highly improbable that the human eye could have evolved by small changes over time. Incredibly, in response to the calculations, biologist Sir Peter Medawar replied that given evolution was a fact, the calculations must be incorrect.[55] Therein lies a fundamental problem. Somewhere along the line, a scientific theory became a religion. When people are no longer open to their theories being tested and potentially rejected on the evidence, there's a name for that. It's called a cult.

Some have been willing to look at the evidence and make rationale conclusions. One example is Dr Ming Wing who graduated from Harvard and MIT and is a world-class cataract and LASIK eye surgeon.[56] After extensive research into the healing properties of the amniotic membrane surrounding babies in the womb, he obtained two U.S patents and invented the first amniotic membrane contact lens. These lenses use the *"...scarless foetal wound-healing property of the amniotic membrane to help in healing the ocular surface of dry eyes..."*[57] The point is, when it comes to the human eye, Dr Wing knows what he is talking about.

Although he grew up an atheist, it was the design of the human eye that led him to firmly believe it could not have happened by random processes, and he concluded it could only have occurred through the careful design of a creator.

For the human eye to have successfully evolved, it required enormous numbers of complex and beneficial mutations. However, the eye does not function in isolation. In addition to the human eye evolving, there needed to be vast numbers of simultaneous mutations across different body systems in a manner that all seamlessly interconnected with the eyeball.

Every single step of this process and every mutation had to be beneficial and not kill off the creature, otherwise, the process would start again from scratch. The problem is that until the eye and all related structures were fully developed and functional, all such mutations were not only unhelpful, but they were dangerous, as they gave the creature a disadvantage in its 'survival of the fittest' struggle.

Bear in mind also, that this accidental 'evolution of the eye' repeated itself across many unrelated species that were all evolving independent of one another. How convenient was that?

Let's consider some of the challenges for the human eye to have evolved by chance. I'm not a scientist, so bear with me. There's no science lingo here. It's just common sense.

———◆●◆———

Two of everything

When describing the process of macro-evolution, the tendency is to simplify the enormity of the task because the truth is a little too scary to confront. Let's not do that. One of the obvious but less discussed problems with the evolution of the human eye, is that it required two of everything to simultaneously evolve. Two retinas, two eyeballs, two eyelids, two eye sockets. Two of everything.

Let that sink in for a minute. Each component of the eye, and its supporting structures such as eye sockets, muscles, and tear ducts were completely unrelated to the other and had to evolve independently in its own timeframe, and yet they are an exact mirror of the other.

Don't forget the eye socket

An eyeball on its own is useless without a perfectly structured eye socket. However, this presents an enormous problem. The eye socket is part of the skeletal structure which is an entirely separate system to the muscular and nervous systems which the eyeball is part of.

As an aside, it's worth noting that all 10 systems of the human body had to evolve independent of each other, but they are all completely dependent on the other to be of any use. But that's another story.

From an evolutionary perspective, there is no relationship between the skeletal system and those systems that formed the eyeball, and yet they had to dance together in perfect evolutionary symmetry for millions of years to arrive at a successful outcome.

We are asked to believe that over millions of years, countless random mistakes occurred in the skeletal structure in a manner that formed not one, but two perfectly suited holes that would just happen to perfectly fit not one but two eyeballs.

In addition, the eyeballs evolved over millions of years, and found themselves perfectly fitting into the unrelated eye sockets. Furthermore, the eyeballs and eye sockets conveniently 'chose' to situate themselves in the best position possible in the human body where sight was maximised, and the distance to the body's computer was minimised. Fortunately, the eyeball didn't evolve and end up on the big toe while the eye socket ended up on the hip.

We are also asked to believe that once the skeleton and eyeballs were working perfectly, the unguided and random mutations 'decided' to stop happening because the job was done.

All of this occurred with no purpose, no plan, no coordination, and no intelligence. This is like giving two separate engineering companies the following brief. One company is given the task of creating a structure that must perfectly contain an object they are not allowed to know anything about, and the margin of error must be zero. The company is not told the purpose of the object or its size or what it is made of.

The other company is given the task of building an object to fit perfectly into the other company's structure, and the margin of error must be zero. They are not told what function the object must serve, nor how big it must be or what material it must be made of.

To make it more realistic, both companies are not allowed to use any tools, and no one is allowed to communicate with each other in any way during the project, either within each company or between each company.

I've met some bright people, but there's no company in the world that would take on a job like that, and yet we are told the eye socket and eyeball came about by accident. Interesting logic wouldn't you say?

Integration with the brain

A fully functioning eyeball is incredible. However, on its own it is useless unless fully integrated to the machine of the brain. That might sound simple but it's not. While the eye was busy for millions of years having constant accidents and mistakes until perfectly formed, the brain had to be very busy as well.

The brain had no concept of light or how to receive it or how to interpret it. It had no concept of an eyeball and no idea what it's function was. Despite these challenges, two significant things had to evolve independently and yet be in critical and perfect alignment with one another.

First, the nervous and muscular systems had to evolve structures we call optic nerves. They had to evolve at the perfect location behind the eyeball, then grow towards the brain and somehow connect with it. Remember, this had to happen perfectly, twice. Each eyeball making the exact same mistakes at the exact same time.

While oblivious to all of this going on, the brain had to independently make lots of mistakes to set aside a part of the brain we now call the visual cortex, so that it could receive electrical impulses from the optical nerves. However, that's only the start of the problem.

Assuming all the mistakes somehow led to each eyeball forming an optic nerve and connecting perfectly to the brain, they were useless unless something more complex also happened over millions of years by accident.

While all these mutations were happening, the brain needed to be very busy making lots of mistakes as well. The optic nerves are 100% dependent on the brain's software perfectly evolving to receive the complex information from the nerves, interpret it, and then send the appropriate signals to all other parts of the brain that make use of the information. Without software, the eyeball is dead in the water.

If you're still wondering about whether this is doable by chance, consider how you plug a USB stick into your computer. Not only does it fit perfectly, but as soon as you plug it in, the computer recognises it, reads the data, and opens a folder so you can see in your language what files are on it. All of this takes software that is specifically designed for this purpose. It doesn't happen randomly. The same is true about the eyeball connecting to the brain. The brain's software reads the images, interprets them, and makes use of them. As it is logical to conclude a designer was responsible for the computer reading the information on the USB stick, so it is also logical to conclude a designer was responsible for the eyeball and the software of the brain.

Eyelids

The difficulties for evolution are not over yet. Assuming for a moment, all the above actually happened by chance, there's so much more that had to happen perfectly.

The eyelid is entirely separate to the eyeball and the skeletal socket, and yet without the eyelid, the eyeball is in trouble. The eyelids protect the eye from danger, including light and debris.

Without even thinking about it, your eyelids stay open automatically during the day and stay closed at night. When you blink, you don't have to focus on blinking one eyelid and then the other. They work in harmony.

The story goes that our eyelids formed by accident over millions of years, and very conveniently, the evolutionary timeframe coincided with the evolution of the eyeball so that the eyeball was quite literally, not left high and dry.

Eye ducts

What about our eye ducts? Without eye ducts your eyes would dry up and die very quickly. They are critical for the eye's survival, but this presents a major problem for evolution.

The eye duct is completely separate to the eyeball, eye socket, eyelids and brain, and yet not one but two eye ducts formed in exactly the right place within each eye socket at exactly the right time. Evolutionary theory says this process occurred by accident over millions of years, and yet arrived at a perfect and absolutely essential outcome.

Eyelashes

A nice set of eyelashes does wonders for how we look, but they play a far more important role than that. On average we have about 320 eyelashes. They protect the eye from dust and foreign objects.

Here's the problem. A set of eyelashes didn't just pop up with one lucky mutation. Each eyelash needed to form on its own, with the right type of hair in the right place, facing out and not into the eyeball.

This mutation needed to happen on the tops of the eyelids and the bottom. Think about that. There needed to be 320 completely random mutations working in perfect harmony.

Each eyelash needed to mutate to become the same size as all the others, perfectly in the right place, with the ability to regrow even when it falls out.

Perceiving light

The evolution of the eye is surrounded by many elephants in the room. It's a crowded room already, but let's introduce one last elephant and that is the question of how the concept of light could in any way relate to random biological mutations.

Let me engage your professional services for a moment. In exchange for paying you an exorbitant fee, I want you to build a machine to receive signals and data from an invisible force that is currently unknown to man. I also want you to then use those signals for a purpose that I won't tell you about. You have the whole world of engineers at your disposal, unlimited funding and no deadline. It sounds impossible right? Well, you've just been introduced to the challenge of the human body evolving a tool to perceive and make use of light.

How could an unintelligent organism understand the concept of light and determine how to build an incredibly complex machine to perceive that light and make use of it?

Some have tried to push the elephant out of the room by arguing that the eye evolved from primitive creatures with 'light sensitive cells'. However, these theories only compound the problem. How did the cells evolve to become light sensitive and how did these cells become aware of something called light? Remember, evolutionary theory says that helpful mutations survive, and unhelpful ones die off. Any creature with a partially developed light sensitive cell would die off as it served no purpose until fully functional.

The same problem applies to how unguided random processes could comprehend what soundwaves are, or taste, or smell. Our noses, tongues, and ears are all highly functional machines specifically dedicated to receiving very specific data from our environment. The idea that millions of accidents led to the evolution of sensory receivers to receive something that was not even known or understood is remarkable.

———•◆•———

Insurmountable problem after problem is explained away by evolutionary biologists with the phrase *"provided there is enough time, anything is possible."*

Perhaps you agree with them. However, I must confess to not having enough faith for that belief. Like Dr Ming Wing, I prefer to form views based on the evidence and when I consider the complexity and function of the eye and all its necessary supporting components, I consider the most obvious and logical conclusion is that it was designed by an extremely intelligent being.

DESIGN ALWAYS COMES FROM A DESIGNER, NO EXCEPTIONS

"The appearance of design proves a designer unless there is substantial evidence to the contrary"

Mathematical design in nature

When I was at school, I developed a liking for mathematics. The reason is with mathematics you are either right or wrong. There are no half right answers. Unlike other subjects such as English, History and Philosophy, Mathematics is strict and unforgiving. It's not something that is made up. It just is because it always has been.

One mathematician explained it this way *"...mathematicians aren't making up what they do. It's not a case of someone with a blank sheet of paper asking what they'd like mathematics to be. Mathematics is already there and has always been – part of the eternal and beautiful order of the mind of God... There is no way that the atheist can account for the fact that the physical world, the mind, and mathematics, make such a perfect fit together"* [58]

Here's the difficulty for atheists and evolutionists. Although our natural world is oblivious to maths classes and calculators, we find that nature seems to follow a grand mathematical architectural code that applies across all its parts no matter how unrelated those parts may appear. [59]

If we are to believe in randomness as the explanation for the universe and our natural surroundings, then there should be no connection between things such as the shape of a shell on the beach and the shape of spiral galaxies, and yet there is with astounding precision.

One such discovery of mathematics in nature is called the 'Fibonacci Sequence' [60]

In the 12th century, Italian mathematician Leonardo Fibonacci observed a series of numbers and patterns which create a spiral when plotted on graph paper. Beginning with zero and one, the sequence continues forever by adding the two previous numbers and creating the next number: 0, 1, 1, 2, 3, 5, 8, 13, 21, 34, 55, 89. When plotted on paper, the sequence creates a spiral as outlined below.

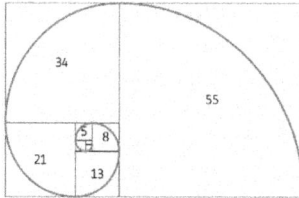

On its own, this numerical series is no big deal. Mathematics has all kinds of formulas that create impressive images when plotted on graph paper. However, this one is particularly intriguing. Not only does this sequence of numbers look fascinating on graph paper, but it is also embedded throughout nature.

Whether it's the inside of your ear, your fingerprint, the shape of the oceans' waves, spiral galaxies, atmospheric storms, and even the rows of petals on flowers. All follow the Fibonacci pattern. We also see it displayed in the ratio of the various joints in our fingers to one another. It gets more interesting when we consider the Fibonacci sequence even has a bearing on how attractive we find each other. Generally, the more aligned our faces are to the Fibonacci sequence, the more attractive we appear.

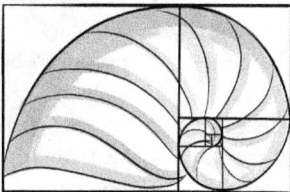

There is no natural explanation for why a fingerprint and a spiral galaxy adhere to the same mathematical properties, and yet they do.

When observing this phenomenon across nature, the logical conclusion is someone designed it this way. There is no other plausible and intellectually honest explanation for how entirely unrelated things follow the same mathematical structure.

The appearance of design proves a designer unless there is substantial evidence to the contrary. The presence of the Fibonacci sequence in numerous unrelated parts of nature only reinforces this conclusion.

We are software driven creatures

When I first started using a computer, it was a lot different to the ones we use today. Back then, the internal memory of a computer was so small, we had to insert floppy disks into the computer which contained the necessary software for the computer to operate. Eventually, floppy disks were replaced with compact discs (CDs). Floppy disks and CDs were hardware. On their own, they were useless, but when software was loaded onto them, they became functional and gave instructions to the computer to perform complex tasks.

As you read this book, you are also using a range of hardware. Your hands are helping hold a book or kindle and your fingers are turning the page. Your eyes are darting across the words and the computer inside your skull is processing everything you are reading. That's all hardware, but just like the floppy disk, unless it has software to run it, your hardware is useless.

Hardware without software gets you nowhere. Because we don't see our software, we don't pause to consider how it got there or how complex it is.

Before we think about the software in our DNA, let's think about something we might be more familiar with. Every day we use computers, whether it's a phone, a PC or tablet, or even our cars and microwaves. They are all examples of hardware that are driven by software. We know with 100% certainty that none of that software came about on its own, and was instead created by intelligent software coders. They were told what the hardware was and its purpose so they could then design and write the detailed coding to run the hardware.

I'm guessing most of us have never looked at computer software. Like the dictionary, it's not the kind of thing you grab off the bookshelf to have a good read with a cup of coffee. If that's the case, then take the following example as the one and only time you will do so.

Computer gaming is a big deal. In 2022, the global revenue was USD $214 billion. Modern games are incredibly complex and enable people all around the world to play the same game at the same time. For some of the games, writing the software is more like making a movie and involves artists, programmers, designers, audio specialists, and producers.

When a game is developed, it starts off with a concept which outlines what the game is about, its characters, scenes, and levels, and how the user will interact with the game. Once that is defined, software developers then begin the task of writing computer code to bring the game to life. The code is then written and tested over and over to make sure the game works. If the code is wrong, it will cause a fault and the game won't operate properly. There's nothing worse than progressing through the levels and then the game freezes!

If you're old enough, you might remember a simple 2D game called 'Brick Out'. It involved a bat sliding up and down the screen to hit a ball which then bounced against a brick wall. Each time a brick was hit, it disappeared, and this was repeated until all the bricks were gone. In terms of a game, it doesn't get simpler than that. However, even this simple game required software and it looked something like this. [61] I don't expect you to read it, but it will give you the general idea.

Game Code

```
import pygame
import sys
from pygame.locals import *
# Initialize Pygame
pygame.init()
# Constants
WIDTH, HEIGHT = 600, 400
BALL_RADIUS = 10
PADDLE_WIDTH, PADDLE_HEIGHT = 100, 10
BRICK_WIDTH, BRICK_HEIGHT = 50, 20
WHITE = (255, 255, 255)
BLUE = (0, 0, 255)
# Create the screen
screen = pygame.display.set_mode((WIDTH, HEIGHT))
pygame.display.set_caption("Brick Breaker")
```

```python
# Create the paddle
paddle = pygame.Rect(WIDTH // 2 - PADDLE_WIDTH // 2, HEIGHT -
PADDLE_HEIGHT - 10, PADDLE_WIDTH, PADDLE_HEIGHT)
# Create the ball
ball = pygame.Rect(WIDTH // 2 - BALL_RADIUS, HEIGHT // 2 - BALL_
RADIUS, BALL_RADIUS * 2, BALL_RADIUS * 2)
ball_speed = [5, 5]
# Create the bricks
bricks = []
for row in range(5):
    for col in range(WIDTH // BRICK_WIDTH):
        brick = pygame.Rect(col * BRICK_WIDTH, row * BRICK_HEIGHT,
BRICK_WIDTH, BRICK_HEIGHT)
        bricks.append(brick)
# Main game loop
while True:
    for event in pygame.event.get():
        if event.type == QUIT:
            pygame.quit()
            sys.exit()
    # Move the paddle
    keys = pygame.key.get_pressed()
    if keys[K_LEFT] and paddle.left > 0:
        paddle.move_ip(-5, 0)
    if keys[K_RIGHT] and paddle.right < WIDTH:
        paddle.move_ip(5, 0)
    # Move the ball
    ball.move_ip(ball_speed)
    # Ball collisions with walls
    if ball.left <= 0 or ball.right >= WIDTH:
        ball_speed[0] = -ball_speed[0]
    if ball.top <= 0:
        ball_speed[1] = -ball_speed[1]
    # Ball collision with paddle
    if ball.colliderect(paddle) and ball_speed[1] > 0:
        ball_speed[1] = -ball_speed[1]
    # Ball collisions with bricks
    for brick in bricks:
        if ball.colliderect(brick):
```

```
    bricks.remove(brick)
    ball_speed[1] = -ball_speed[1]
# Draw everything
screen.fill(WHITE)
pygame.draw.ellipse(screen, BLUE, ball)
pygame.draw.rect(screen, BLUE, paddle)
for brick in bricks:
    pygame.draw.rect(screen, BLUE, brick)
# Update the display
pygame.display.flip()
# Control the frame rate
pygame.time.Clock().tick(60)
```

If you actually read any of that, you must be very intrigued or very bored! In the world of gaming software, this is basic coding. However, if merely one word, symbol or number was incorrect then the game wouldn't work properly.

———◆◆◆———

No intelligent person would conclude this computer code came about by someone randomly hitting the keyboard with a blindfold on. Why not? It's simple. The computer game has a function or purpose, and we know that design always comes from a designer, no exceptions.

You know where this line of thought is going. If we conclude a simple computer game has a purpose and therefore the software was designed by a designer, the logic must also follow when we look at the natural world. If we see something in nature that has a useful function or purpose, we should logically conclude a designer was responsible, not only for the design but for the underlying software. That is the most rationale conclusion.

Let's think about this a bit further. While computer software is complex and requires careful design, it looks like child's play when compared to the software inside just one biological human cell. On average, a male has approximately 36 trillion of them.

Contrary to what we were taught in school, it turns out the 'simple cell' is more complex than any computer software ever written. Not only does it comprise intricate coding, but it also contains microscopic machines with spinning motors, all with their own purpose and function, each doing their job to protect and replicate the cell.

Some have compared the inner workings of one 'simple cell' to the functioning and structure of a small town.

———————◆•◆———————

In 1953, American biologist James Watson and English physicist Francis Crick expanded our understanding of DNA. They discovered that the DNA molecule consists of a three-dimensional double-helix. You might ask so what? Well, more than a fancy structure, DNA comprises genetic information and instructions for the functioning and growth of all organisms including you.

DNA has been described as a chain of individual molecules known as nucleotides. There are four nucleotides - Adenine (A), Thymidine (T), Guanine (G), and Cytosine (C). In essence, these four nucleotides are like letters, and the way the letters are put together throughout a strand of DNA creates software, just like a computer code. The code then dictates what cells are formed and what they do.

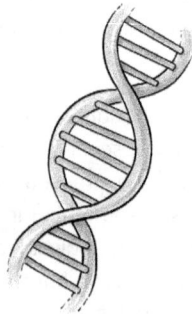

Every day, your body replicates it's cells, repairing some and replacing others, and it just happens because that's the way you are coded. The skin on your face is different to the skin on your fingers because the software code is different for each part of your body. It all happens without thinking about it because your personal DNA software is pre-programmed and just goes about doing what it was designed to do.

It is difficult to comprehend how complex our DNA software is.

If our DNA was laid out in a straight line, it would form one long word of approximately 3.2 billion letters. That would equate to an A4 book of approximately 853,000 pages. Stacked one on top of each other, the books would reach about 100 metres into the air. Can you get your head around that? That's what your individual DNA looks like. But remember, the combination of those four letters across the 3.2 billion letter word is not random. It can't be because any errors in your software would be disastrous.

Pause for a moment and ask yourself, do you really believe that your DNA of 3.2 billion perfectly sequenced letters, reaching 100 metres into the air came about by randomness? If the answer is yes, can you see what you just did? On the one hand, you wouldn't conclude that a simple computer code could come about without a software coder, but when it comes to our DNA, the most complex coding in the universe, you are happy to bypass logic and conclude it was the result of chance. Why is that?

Being confronted with the facts is confronting indeed. It forces us to wrestle with the logical consistency of our conclusions. I believe if we assess the evidence objectively and unemotionally, the conclusion is obvious and unavoidable.

WHAT'S YOUR CONCLUSION?

"People almost invariably arrive at their beliefs not on the basis of proof but on the basis of what they find attractive."
(Blaise Pascal)

The classroom says we came from nothing, for nothing, to nothing. It tells us the most complicated software code in the universe came about with no intelligence or design. It says the one conclusion we are not allowed to make is that there was perhaps an intelligent God behind it all.

We are told to ignore in nature the 'appearance' of design, purpose, and complexity. Instead, when we look at the stars and contemplate our purpose and existence, we are to conclude that such thoughts are merely the result of evolutionary misfiring in our brains.

Earlier, we mentioned C.S Lewis' quote *"If I find in myself desires which nothing in this world can satisfy, the only logical explanation is that I was made for another world"* [62] However, because our existence is accidental, we are encouraged to bury such thoughts, and when they pop up, we must remember back to the primordial soup we came from and be thankful for the fortuitous bolt of lightning that kicked this whole thing off.

———◆◆◆———

Contrary to classroom ideology, the Bible gives us a very different account of how and why we exist. It lines up perfectly with the evidence we see in nature and in the fossil record, and it addresses the inner questions and desires we all have.

In the Bible, Psalm 139 tells us we were created by God on purpose, and for a purpose. It says *"For you created my inmost being; you knit me together in my mother's womb. I praise you because I am fearfully and wonderfully made; your works are wonderful, I know that full well. My frame was not hidden from you when I was made in the secret place. When I was woven together in the depths of the earth, your eyes saw my unformed body. All the days ordained for me were written in your book before one of them came to be."*

The book of Ecclesiastes (chapter 3, verse 11) also says that *"God has set eternity in the hearts of men"*. That explains why we feel there's something more.

The Bible tells us we have questions of eternity because we were made by an eternal God. We were created not only as physical beings but also as spiritual beings with an eternity ahead of us. Now that's a very different story, and its one worth exploring.

———————◆●●————————

Perhaps you've read this book so far and have concluded that despite the evidence, it is more likely than not that all we see and know is one big accident of astronomical probabilities built one on top of another. That's a conclusion you are entitled to make. I personally don't have enough faith to believe that.

However, we should take note that the Bible asserts the evidence for God's existence is so strong that it says in Psalm 14, verse 1 *"The Fool says in his heart 'there is no God'"*. That's not trying to offend people. What it means is that if after considering the evidence, you conclude there is no God, such a conclusion according to the Bible, is not a rationale conclusion of the mind, rather it is a conclusion of the heart. This means that some people simply do not want to believe in God.

When we dig a little deeper, we often find people have other reasons for not wanting to believe in God. Perhaps they were deeply hurt in their life or suffered loss and they thought God should have looked out for them and he didn't.

Some don't want to believe in a God who let them down, and others say they can't believe in a God who allows suffering in the world. Those are very strong reasons. The hurt is real, and it should not be explained away in some trivial manner. However, those conclusions are ones of the heart and not the mind. They come from a place of emotion and not evidence or logic.

There is no logical connection between the existence of suffering and the existence of God. To conclude there is no God because there is suffering means firstly, we have already defined who God would be if he did exist, and secondly, we have decided it was him who caused or at least permitted the suffering. It doesn't work like that. If God exists, he is who he is irrespective of the existence of suffering or evil in the world.

In addition, if we conclude that the presence of evil or suffering means God does not exist, then what does the existence of goodness, joy and love lead us to believe? We can't have it both ways.

Death and suffering are realities in our world. However, if evolution is true, then the existence of suffering should be accepted without complaint. We should have no objection because death and suffering are merely the product of random purposeless processes. In fact, they are a necessary requirement of evolution.

However, the truth is we know intrinsically that things are not the way they ought to be, and we feel aggrieved when death and suffering occur. This universal objection to death and suffering is not only inconsistent with the doctrine of evolution, but it also confirms to us that we are meant for something more, and that the answers are not found in a bowl of primordial soup.

We might not understand how to reconcile the presence of suffering with the existence of God, but we also cannot reconcile our objection to suffering with our belief in evolution, and so we are faced with a choice of what to do with these competing ideas.

In weighing up the evidence for the existence for God, I consider the most logical conclusion is there is a God despite the presence of suffering, and the real question then is how this can be so. Perhaps it is in this seeming contradiction that we are about to discover the greatest truth of all.

———◆◆———

Maybe you are at the other end of the spectrum and have formed the view that it is more logical to conclude there is a God who created us. If so, I take my hat off to you because I personally consider it's the only logical conclusion. However, if that is your conclusion, the next questions are why does it matter, and if there is a God, what's his phone number?

As the theologian Professor Van Til once said, *"One does not automatically stand in awe of God's creation and then in doing so know of God's will, of sin and of pardon"* [63] In other words, it's one thing to conclude God exists but that doesn't mean you know how to contact him.

HOW CAN WE KNOW GOD?

"Most of us know about God, but that is quite different from knowing God" (Billy Graham)

In 2022, a Gallup poll found 81% of Americans believe in God. For most people, it seems that believing in God is obvious and they don't need to be convinced. However, the next question is who is God? There seems to be so many to choose from.

Some would say the God you believe in is strongly influenced by the environment you were born into, and the religion of your culture and family. If that's the case, then why should we believe one view of God over another?

In 2020 there were 1.8 billion Muslims in the world and 2.38 billion Christians. If only one view of God is right, can we seriously say that over a billion people are wasting their time worshipping a figment of their imagination? That would seem rather arrogant, or intolerant wouldn't you say?

The marketplace of religion is indeed a very noisy place with all the traders yelling out their beliefs, offering what they consider is the truth. It reminds me of when I walked down the bustling streets of Thailand with all the stall traders trying to sell me their goods. The religious noise is overwhelming, and understandably for some, they prefer to leave the traders to argue amongst themselves.

———◆•◆———

Let's consider for a moment the vast number of gods in the religious marketplace. You will be aware of some of the more familiar ones such as Jehovah, Jesus Christ, Allah, and Buddha. But that is just the start.

Growing up in India you would have 33 million gods to choose from.

You may have seen images of their most revered god which looks like a human with an elephant's head.

In ancient Greece or Rome, there were plenty of gods to pick from. Zeus the god of the sky, lightning and thunder, Neptune the god of the sea, Mars the god of war, Apollo the god of the Sun, music and prophecy, Artemis the god of nature and the moon, Vulcan the god of fire, and Aphrodite the god of love. Even today, our Marvel and DC movies have reintroduced us to these gods and have brought them to life on screen.

In ancient Egypt you would have worshipped the gods Osiris and Horus. Even today, for some reason, the American $1 dollar note has a picture of an Egyptian pyramid with the eye of Horus hovering over the pyramid. Whatever the reason, religious beliefs have continued to permeate society and government far deeper than many appreciate.

Being raised in an Arab nation, you would likely have worshipped Allah, a god introduced in approximately 550AD by a man named Muhammad.

Followers of Buddhism worship a range of deities and believe that reincarnation repeats itself until you reach a state of perfection.

There are even groups who worship alien races from other planets, believing they are humanity's creator and are returning soon to save the world from self-destruction.

If you were born a Jew, it is likely you would follow Judaism which believes in one God named Jehovah who created the universe and created mankind in his image. A close cousin to Judaism is Christianity. As a Christian you would also believe in Jehovah, but in addition would believe God came to earth in human form in the person of Jesus Christ who then died on a Roman cross for our sins and rose again, and now offers eternal life to everyone who believes in him.

With so many different religious beliefs, is it possible to weed through the noise and discover the true God, or can we only conclude there is most likely a God and leave it there?

Before the internet, DVDs, CDs, and cassette tapes, we only had record players and radios to listen to. I'm old enough to remember those days, and I recall often turning the dial on the radio looking for my favourite station. Often as the dial moved, I would pick up a lot of indecipherable static. At times, I would also hear what appeared to be two stations coming through the radio at the same time. I could pick up the odd word, but it was very confusing.

The trick was to slowly turn the dial until it eliminated the noise and the needle settled on one particular station. When that happened, the signal came through loud and clear.

Just like the radio transmission, as we turn the dial of religion, there is a lot of static and noise. However, there is one frequency that stands out from all the others. It makes such bold and exclusive claims regarding truth that we can listen in and decide whether it is completely true or completely false. Unsurprisingly, the religion I am referring to is Christianity and it centres on a person called Jesus Christ.

Here's a confronting thought. If Christianity and the claims it makes about God are true, then every other religion and view of God is false, irrespective of how many people follow those other religions. Christianity is utterly exclusive and does not give the option of merging with other religious frequencies.

That might not sound politically correct, and it's not. A genuine search for truth is only genuine if it discards things that are untrue, irrespective of how many people may be offended. I cannot believe 3*3 = 9, while also accepting 3*3 = 7 or 8 because I don't want to hurt peoples' feelings. That approach to truth would not have helped the study of mathematics. Once we conclude what truth is in regard to God, we should be willing to stand upon that truth and discard all other beliefs, while at the same time showing love and respect for those who do not share our belief.

Exploring Christianity is a very helpful place to start in a search for God because it immediately narrows down the question of who God is.

If you can conclude Christianity is true, then the search is over. Conversely, if it's not true then you can dismiss it and continue your search.

———◆◆◆———

Approximately 2000 years ago, something happened in the Roman Empire that changed the world forever. A man from an insignificant town and upbringing, with no formal education, appeared on the scene and started teaching people about God. At 30 years of age, he challenged and confounded the religious leaders of his time. He shared a simple message that God the creator loves us and wants us to love him back.

The people said he spoke with love and authority unlike the religious leaders they were used to, but in addition he demonstrated his authority by performing miracles, healing the sick and driving out evil spirits.

While many people loved him, there were many religious leaders who hated him because he broke their system of power and control. He didn't follow their rules. To deal with him, they had one solution. Working with the Romans, they orchestrated his crucifixion and Jesus was nailed to a cross and died. It seems the cancel culture was also operating in his day.

The existence of Jesus and his crucifixion is a fact of history. There is no serious history scholar who disputes it, and there is a wealth of ancient non-biblical texts that affirm it. However, the most significant part of the story is that the Bible claims Jesus not only died, but he also rose from the dead. That's either nonsense or it's the most significant event in history and has enormous ramifications for us all. If it's true, then it tells us Jesus was a lot more than a good moral teacher.

Jesus himself told his disciples on numerous occasions that he would be crucified and that after 3 days he would be raised to life. If it did not happen, then quite frankly, Jesus was a false teacher and not to be trusted.

Whether or not Jesus rose from the dead, there is an abundance of evidence demonstrating that many people alive at the time of Jesus genuinely believed he did, and they were willing to die for their belief.

Even with such a cursory glance at this story, there is too much here to merely dismiss without further examination, so let's go on and do so.

The life of Jesus was so significant that it changed our calendar. I am writing this book in the year 2024 AD. AD means Ano Domini, a Latin phrase meaning 'in the year of our Lord' which is a reference to Jesus. Whatever view we hold about who Jesus really was, we can't get around the fact we are surrounded by his legacy.

His name is even used as a swearword on building sites, in offices, schools, and homes. We may not all know him, but we certainly talk a lot about him!

Other religions do not know what to do with Jesus. They can't deny he was a good moral teacher, and yet they can't completely accept his message because his message nullifies their religion. It does so because Jesus taught there was only one God, and that he was the only way to know God. If he is correct, it logically follows that all other religions are redundant.

The claims of Jesus are so non-negotiable and exclusionary that it is appropriate to demand significant evidence in order to accept them.

So far, in this book, we have examined a range of evidence in the natural world and asked whether it points to the existence of God. Consistent with that examination, if Christianity and the God of the Bible are true, then the Bible should align with that evidence. With that in mind, let's assess some essentials of Christianity to determine whether it holds water.

What does Christianity say about how life commenced?
At the start of this book, we established that science and the Bible agree the universe had a beginning, but there is no consensus on what caused it to begin. The only two options are either the universe was caused by nothing, or it was caused by God.

The Bible's explanation is simple, *"In the beginning, God created the heavens and the earth."*

According to the Bible, with unfathomable power, God spoke the universe into existence. He didn't need raw materials to create the universe. He simply declared it to be. The book of Hebrews (chapter 1, verse 3) describes it like this *"...the universe was formed at God's command, so that what is seen was not made out of what was visible"*. Even back when the Bible was written, people were contemplating the concept of what caused the universe to exist. While the atheist must believe the universe spontaneously arose from nothing, the Bible says it was God who was the first cause of the universe.

Science has no answer to how the universe came from nothing, and it has no answer to how life came from non-life, but the Bible has an explanation for both. It says that once the universe was formed, God was personally responsible for the commencement of life. He was the master physicist, mathematician, and software coder. It was he who wrote the 3 billion letters of human DNA and the DNA for every living creature.

The Bible provides no scope for accepting a theory of gradual mutations over millions of years ultimately resulting in humans and other species. Instead, the book of Genesis states that God made all types of creatures *"according to their kind"*. This means all species were fully complete from day one. There was no process of pond scum evolving into fish, or fish into birds. By God's eternal creative power, he made all species perfect *"according to their kind"* with no prior forms.

When discussing these topics, terminology can cause confusion. For the purpose of our discussion, we are using the words 'kind' and 'species' interchangeably to mean the range of creatures that exist. Dogs and cats are 'species' or 'kinds' of creatures, and there are large varieties of those creatures.

If the Bible is correct and creatures were fully formed *"according to their kind"*, then we would expect to see micro-evolution regularly occurring throughout successive generations of creatures. Equally, we should not see evidence of macro-evolution, namely one kind of creature evolving into another. If we did, the Bible's account would not be right.

The reason we should observe micro-evolution is because each original 'kind' of creature contained all the necessary DNA to enable numerous variations to occur with each successive generation.

Whether its horses, cats or dogs, the original kind of each creature contained all the necessary genetic code to branch off into different variations of horses, cats, and dogs. However, the key thing to note is while different breeds of dogs combine and form a different looking dog, the offspring is still a dog. This is micro-evolution at work, and it's exactly what we would expect from the Bible's view of creation. These changes are not evidence of macro-evolution. There is no changing from a dog to a horse. We don't see macro-evolution in the fossil record and we don't see it happening in living creatures.

The creatures we see alive today and those in the fossil record, are entirely consistent with the Biblical account of God making creatures *"according to their kind"*.

The conclusion appears obvious. Not only does the evidence in the fossil record and in living creatures substantiate the Biblical account of creation, but it also undermines, or some would say disproves the evolutionary account.

———————•••———————

What does Christianity say about mankind?
According to the Bible, mankind is the pinnacle of all God's creation and has been given authority to rule over it and to operate as a caretaker on God's behalf.

We are not God, but we were made in his image, and this sets us apart from the rest of creation. Not only are we flesh and blood like other creatures, but we are also spirit, just as God is spirit.

God made us for a specific purpose and when we live in that purpose, we find meaning, fulfilment, and life. That purpose is defined as knowing and having a relationship with our creator and then living out our life from that perspective. Without knowing God, we can still find meaning and fulfilment, but not in the way God intended and not in a manner that continues into eternity.

The book of Genesis says when God made the first humans, Adam and Eve, he placed them in what must have been an exquisite part of his creation, known as the garden of Eden. God appointed them as gardeners and caretakers of what he had made.

In this environment, Adam and Eve had an intimate and meaningful relationship with God. We read in Genesis that Adam and Eve *"heard the sound of the Lord God as he walked in the garden in the cool of the day"*. God taking a walk with the people he made. That's amazing. I wonder what that sounded like and whether God was singing as he walked through the garden.

God always intended to have an intimate relationship with us, and he still desires to have it with you and me.

Now, I hear you asking, if God so desperately wants that type of relationship, why don't we see him or hear him walking near us? Why does he seem so distant? I understand those questions. They are the natural ones to ask because they are real, and they need real answers, and we will address those shortly.

To get there, we need to understand that Adam and Eve experienced a unique time in history that we don't get to see today. While we live in a broken world, they did not. They were privileged to experience a world that had yet to know death and suffering. We can't imagine a world without those things because it's all we know.

You may come across a pile of shattered pottery that was once an exquisite art piece. You will never know what it looked like before it was broken. Only the person who made it will know. It's the same with the world we live in. Its broken, and it's all we know.

———◆●◆———

If creation was originally perfect, then what happened? The world we live in today is beautiful in many respects, but it's also full of brokenness, sickness, and death. It's a world where God seems distant. Why is that?

We all know something is wrong with this world, but we can't put our finger on it. We don't merrily embrace suffering and death, because we sense things are not the way they ought to be. It's like looking at the broken pottery vessel and knowing it was not created broken.

93

It was originally perfect and was meant for something more, but we just don't know what that is.

It's not as if we don't try to fix our brokenness. Mankind invests billions of dollars every year attempting to extend life, fight disease, heal depression and solve poverty. We do so because we share a collective sense that death and suffering are wrong and should be eliminated.

So, how did things end up this way? Why are things not the way God intended them? Why is the pottery so broken? The Bible explains that the brokenness we experience is partly to do with God and partly to do with us.

When God created humans, he wanted a genuine loving relationship, built not on compulsion but on choice. To have that type of relationship, God had to take a huge risk. As the creator, he could have made humans as relational robots who had no choice other than to love him, but that would not have been a genuine loving relationship.

True love only exists where it is optional and stems from freewill. True love is far more than a feeling. It's the kind of love that puts others before ourselves. That kind of love goes the distance. God extended that kind of love to us. It's the kind of love God wanted from mankind, and to have it, he needed to create humans not only with the choice to love him, but also the choice not to.

Just like the honeymoon period in a marriage, the first humans were content, happy and in love with God. How could they not be? God provided everything they needed, and he engaged with them every day in a loving purposeful relationship.

It's hard for us to imagine, but as they were created perfect, they had no understanding of evil, guilt, wrongdoing, suffering, sickness, or death. Unlike us, they started life with a perfect relationship with God. They were either going to remain in that relationship or fall away from it. We on the other hand, start our life from a perspective of having no relationship with God and having no idea how to fix it.

———— •◆• ————

How do you give perfect people living in a perfect world, a tangible way to exercise their choice to either love or not love God? This was

God's dilemma. To achieve it, he asked Adam and Eve to enjoy all the food and abundance he had given them, but to avoid the fruit from just one tree in the garden he had placed them in. It was a simple command, and one that would demonstrate their loving commitment to him. God also warned them that if they ate the fruit, they would experience death, both physical and spiritual.

This was no ordinary tree. Not only was it a symbol of choice, but in tasting its fruit, a metamorphosis would occur deep within their soul that would change humanity's nature forever. Adam and Eve would suddenly understand the difference between good and evil, and mankind would struggle between these two extremes through all generations.

———◆●◆———

We all like to buy new things. There's nothing like opening that well packaged box and taking out your new TV or phone. Eventually though, that sense of newness turns to familiarity, and eventually to dissatisfaction.

Incredibly, while Adam and Eve lived in a perfect environment, they too became familiar with all they had, and familiar with the closeness of relationship they enjoyed with God.

Questions formed in their minds. Perhaps God was holding back on them. What if the fruit was not a mechanism for death, but was instead a way to make them even more like God?

Instead of talking to God about their questions, they let their thoughts grow. If we entertain and feed a thought too long, it eventually leads to action, and that is what happened with Adam and Eve. While they had the whole land of Eden to dwell in, they felt drawn to the one thing they should not have, and found themselves standing there staring at it, wondering what if?

Eve ate first and then handed the fruit to Adam. They had enjoyed the taste of so many types of fruit, but this time it was different. After the initial tastebud experience, something entirely different happened. From deep within their soul, they knew a change had occurred. They could feel it. Something had died. For the first time, they felt shame and guilt, and a sense of distance from God. Confusion and fear filled their minds. God's presence had left them. They couldn't explain these new feelings. Unbeknown to them, they had just committed the first sin against God and were now experiencing the consequence of death that God had warned them of.

Having sinned against God, Adam and Eve now had within them a sinful nature, and they started desiring things that would take them further and further away from their relationship with God. We know those desires. We have them at times and see them in others. They include immorality, lust, greed, jealousy, selfishness, envy, and anger.

Adam and Eve's relationship with God was severed and an impassable chasm between God and mankind was formed. Mankind had changed but God had not. God remained perfect, holy, and loving, but mankind, now being sinful, could no longer be in his presence.

Sadly, the consequences didn't stop there. As mankind had been appointed by God as head over all creation, when they sinned, they brought death not only on themselves but on all creation. As their bodies began to die, the rest of creation was also cursed and began experiencing decay, sickness, and death.

This broken relationship with God introduced a major problem we are now part of. As descendants of Adam and Eve, we have inherited the consequence of their decision to disobey and walk away from God. We also have a sinful nature, and we also are born out of relationship with God.

When we are conceived, our parents can only pass onto us the biological traits they have. That's why we look like our parents, and we share parts of their personality. This is the case biologically, and it is the same spiritually. When Adam and Eve had children, they could only pass on the nature they themselves had, and that was a sinful nature out of relationship with God. We may not like that. We may object to it and say it is not fair, but it is the reality. Just as we may have wished our parents passed onto us genes that made us taller or better looking, we get what we get, and it's the same with our spiritual nature. We are born with a nature that is sinful and which is out of relationship with God our creator.

The consequences are not only relevant to our life on earth. If we die in a state of broken relationship, we also enter eternity separated from God.

What does Christianity say about how we can know God?

Being born out of relationship with God and having no idea how to fix it sounds rather depressing, and it is, but it hasn't stopped mankind from trying to solve this problem. The quest to reach God is the foundation of every religion, and mankind has tried to do so for thousands of years in all sorts of ways.

Other than Christianity, all religions agree that to fix our relationship with God, we must work hard to demonstrate our sincerity. These religions believe God is distant and angry and maybe, just maybe he will pay attention if he thinks we are working hard enough to show him we mean business.

Working hard to get to God comes in many forms depending on the religion. Brownie points are earned in ways such as attending a church service, performing a ritual, repeating certain prayers over and over, facing the right direction when praying, confessing sins to Priests, fasting, praying to dead saints or angels, going door to door telling people about our religious beliefs, and many other practices. Some religions even believe if we harm ourselves enough, it will remove God's anger. Still, others consider living in isolation will show one's sincerity to God by removing all the worldly distractions. With all these things, the hope is that we will be accepted by God when he weighs up our 'good deeds' against the things we would rather forget in our lives.

I recall talking to two men who came knocking at my door to tell me about their religion. For them, going door to door was a necessary step to be accepted by God. I appreciated their sincerity, but I asked them a question, and this is what it was *'You are on your death bed with 20 minutes to live. Are you certain you have done enough to be right with God?'* They looked at each other and admitted they could not know. That is where religion gets us when it's based on earning our way to God. It ultimately leads us to our death bed with the lingering question 'have I done enough?'

———————◆●◆———————

The message of Christianity tells quite a different story and stands alone in the marketplace of religion. It agrees our relationship with God is broken, but it declares there is nothing we can do to fix it. No good deed or ritual will bridge the wide chasm between us and God. Christianity says God is holy and we are not, and we can't become holy by trying to be nice people. We have little understanding of what it means to say God is holy. He is above and beyond the purest thing we could possibly imagine, and to be blunt, we just aren't holy, and deep down we all know that to be true.

God is not interested in the things we think will earn his favour, because they don't fix the relationship. If one marriage partner has an affair, they can't fix it by giving the other person gifts. Something fundamental has been broken and gifts just don't cut it.

The Bible addresses this point directly and presents a grim picture. It says in the book of Isaiah (chapter 64, verse 6) "All of us have become like one who is unclean, and all our righteous acts are like filthy rags". Long ago, people were called 'unclean' if they had leprosy. Such people were required to call out the word 'unclean' so that others would keep away and not become infected. They were isolated from society. What a sad existence. They hadn't asked to have leprosy. It just was what it was, and they were deemed unclean. That's how we are in front of God. Because we have a sinful nature, we are unclean and there's nothing we can do to fix it. To make matters worse, all the good deeds we think are 'righteous acts' don't make us clean before God.

Don't get me wrong, there are lots of people we would consider good people. But that's not the point. If you are in your garage fixing a car, up to your elbows in grease and you walk to the clothesline to help take down a load of white shirts, you might have had the right intention, but you've spoiled it. Your good deed was like a filthy rag. The fact is we are unclean because we have a sinful nature. We are full of grease, and we can't clean it off.

The Bible goes on and explains this problem further. In the book of

Romans (chapter 3, verse 23) it reads *"...for all have sinned and fall short of the glory of God..."*. There it is. No matter what we try to do to make things right with God, we fall short. His standard is too high and he's too holy. It's like trying to throw a rock across a canyon. We can give it a go, but we fall short every time.

I know, so far it sounds fairly gloomy, but hang in there as the good part is yet to come.

If what the Bible says is true, then every religion that says you can earn your way to God has got it fundamentally wrong. That's confronting. Some would even say its arrogant, but the Bible doesn't play around, and it certainly doesn't give scope to believe that all roads lead to God.

Jesus was very blunt about this. In the book of John (chapter 14, verse 6) he said *"I am the way, and the truth and the life. No one comes to the Father [God] except through me."* That is an exclusionary statement. Jesus was saying there is no other God, no other system, no other religion, no other truth, and no other way to know God except through him. That is politically incorrect, and no doubt would breach the anti-discrimination laws of today.

Jesus was either entirely right or entirely wrong. We can't have it both ways. We can't say Jesus was a good man and a good moral teacher, but then say he was lying about being the only way to God.

C.S. Lewis dealt with this statement from Jesus and said there is one of three conclusions we can make. Jesus was either lying, or he was insane, or he was telling the truth. A simple reading of the Bible reveals that Jesus was not a liar and he was far too profound to be insane. If he was telling the truth, then we are faced with putting aside modern-day political correctness and accepting what he said. This means not only accepting Jesus, but also rejecting all other religious notions and ideas. Today's culture would not take kindly to such a conclusion.

Our age is called the 'post-truth' age where truth is whatever you define it to be. What's true for you is true for you and who am I to disagree? In other words, for example, the Bible might be right in saying Jesus is the Son of God, and Islam's book the Koran might also be right when it says God has no son. The post-truth age says both could be true because truth is what you define it to be.

This is a dangerous path for society to travel, and the faster the train travels, the harder it is to stop. The natural progression of such ideology leads to a society void of rationality with everyone too scared to point out the obvious. Even in the town I live in, some school children have identified as cats and have been permitted to behave as such in the classroom while others are attempting to obtain an education. That's where post-truth leads us.

As an ideology, 'post-truth' falls apart with the simplest of examinations. Indeed, the 'post-truth' age could also be called the 'age of nonsense' and is hopefully one that future generations look back on with astonishment.

To be clear, there is a place for subjective truth. You and I might disagree whether oysters taste nice and both views are subjectively true, but if we disagree on whether oysters exist, one of us is wrong. The objective truth is oysters exist, whereas the subjective truth is I may like them, and you may not.

Contrary to modern ideology, there is objective truth, and it has nothing to do with what we think or feel, or what we want to be true. A child is not a cat, the earth is not flat, and Santa Claus is not real.

There is objective truth, and it necessitates the rejection of those things that are not true. The square root of 9 is 3, and any other answer is wrong. Telling the examiner that for you, the square root is 4 will do you no good.

•••

Where has the post-truth ideology come from and why do we have such pressure to accept it? In short, the post-truth ideology has arisen from society elevating personal rights, and progressively diminishing personal accountability.

In a democratic society, we hold dear the value of individual liberty. We expect our elected officials to promote and defend it. We encode it in our laws and defend it in our courts. We protest when there is a hint of such liberties being removed. Traditionally, individual liberty has meant not only the protection of personal rights, but the expectation of personal accountability. However, our post-truth age has arisen because over time, society has elevated one side of the equation, namely individual rights, while on the other side, it has diminished personal accountability.

Even in a democratic society, protection of a person's rights has had limits. Those limits arise where the exercise of one's rights negatively impacts another person's rights, or damages the rights of society as a collective. In the world wars, individuals could exercise their right to be conscientious objectors, but because the exercise of such rights impacted the safety of society overall, they were imprisoned for the duration of the war. In that scenario, the obligation of the person as a citizen outweighed the rights they had as an individual, and there were consequences for not fulfilling the obligation they had to their fellow citizens.

The post-truth age has elevated the rights of individuals far above the obligations they have traditionally had as citizens of a well-functioning society. The post-truth age operates on the premise of *'do whatever's good or true for you'*. Where's the sense of duty to our fellow citizens in that statement?

The post-truth age has not only elevated the rights of individuals to be whoever they want to be, but it has also removed the right of others to express their own view that such people may be wrong. Even in my country of New Zealand, I now have a legislative obligation to accept and not question the choice of individuals in expressing things such as gender, even if I consider it to be harmful. Questioning such choices comes under the umbrella of 'hate speech' and can result in fines and in some cases imprisonment.

It is not healthy for a society to feed the imagination of children who exercise their right to identify as cats. It is even less healthy to remove the right of the remaining sane members of society to call out the obvious nonsense and declare such a belief to be false. And yet, we see commonsense cowering under the cancel culture, too afraid to call it for what it is.

You might be wondering why I have taken such a tangent in our conversation and raised matters that readers might have opposing views on. However, it has been a necessary step because due to the age we live in, we have ingrained within us a nervousness to call out things that are untrue, and an even greater nervousness to declare that perhaps we are right while someone else is wrong.

We need to confront this inner hesitation because we cannot face the claims of Jesus without being prepared to do just that.

As we have already mentioned, the claims of Jesus are very confronting. They assert that he is right, and every other religion is wrong. There is no scope for believing the claims of Jesus are true while at the same time believing the claims of other religions are equally true.

A post-truth response to the claims of Jesus would judge him as unloving for not accepting others' views. How dare he say that other religions are not right. Who does he think he is? Everyone should have the right to believe what they want to believe without being told they might be wrong. If Jesus were around today, he would most certainly suffer the wrath of the cancel culture, much like he did 2000 years ago.

That may even be your view of Jesus, but let's assume for a moment that Jesus could potentially be right. Let's be brave enough to park all other views for the time being and hold off our conclusion. The next question then, is what did Jesus mean when he said *"I am the way, and the truth and the life. No one comes to the Father except through me."*

———◆●◆———

The answer to our question lies in the reality that we could not do anything to fix our relationship with God. We needed help. If we couldn't make things right with God, then the only solution was for God to make things right with us.

God could have remained distant from our problems and our brokenness, but instead, he came to earth and became a human like us. His name was Jesus which means 'he shall save us from our sins'. Jesus was much more than a good moral teacher. He is God in human form. He walked with us, ate with us, laughed with us, and cried with us. He experienced the joys and the sorrows of humanity. Because he did so, he can relate to you and me with all the struggles we go through. What greater demonstration of love could there be? God came to save those who rejected him by becoming one of them, and he took upon himself the brokenness humanity had caused.

At the beginning of this book, we read a quote from Apollo 15 Astronaut James Irwin who said *"God walking on earth is more important than man walking on the moon"*. He was right about that. God did not keep his distance. He came to save us and restore our broken relationship. If that is true, then the question is how did God restore the relationship, and what is our part in it?

———◆◆◆———

As amazing as it is, God walking on earth would not in itself restore the relationship. Mankind had sinned against God and a debt had to be paid.

When someone breaks the law and steals or commits murder, we expect they will go to court and pay a penalty. We would be affronted if the judge merely let them off with no consequences.

In the sight of God, we also are law breakers. Not only do we have a sinful nature, but we commit sin every day. While imprinted within us is a sense of right and wrong, our sinful nature often steers us toward doing those things we know to be wrong, and in doing so we break God's righteous standards.

So, we are law breakers. If God let us off the hook, then he would not be a just God. It would seem there was a stalemate. We had fallen out of relationship with God, and owed a debt we could not pay, and yet God being just, could not let us off.

It was precisely this hopeless situation that God came to fix himself. It would take a lot more than money to pay our debt. Something far more valuable was needed. In the Bible, the book of 1 Peter (chapter 1, verses 18-19) explains it like this *"...it was not with perishable things such as silver or gold that you were redeemed from the empty way of life handed down to you from your forefathers, but with the precious blood of Christ..."*

God had told Adam and Eve that the penalty for their sin was death. We know that to be true because they died, and we will too. But more than that, the penalty of death carried on into eternity with eternal separation from God.

Death and separation from God is the ultimate penalty, and for us to be rescued, it required someone to take upon themselves that penalty on our behalf. The price could only be paid by someone who was perfect in God's sight and so was not personally subject to the penalty. But therein lies the problem. None of us are perfect, so no one is suitable to pay the price.

Entering this hopeless situation came Jesus, fully God and yet fully man. He had never experienced sin before, but as he willingly died on a Roman cross, God placed all our sin upon him, so that he could be our representative and pay the penalty on our behalf. Speaking of Jesus, the book of 2 Corinthians (chapter 5, verse 21) explains *"God made him who knew no sin to be sin for us..."*

That dealt with our sin, but in addition, as Jesus hung on the cross, he cried out *"My God, my God, why have you forsaken me?"* (Matthew chapter 27, verse 46). The penalty for sin is separation from God for all eternity. On our behalf, Jesus experienced what it meant to be separated from God.

One of the shortest and yet most profound statements ever made was the last thing Jesus said before he died. The book of John (chapter 19, verse 30) records him saying *"It is finished"*, and then Jesus bowed his head and died. Think about that. A dilemma that mankind had tried to fix for thousands of years was finally resolved on a wooden cross as the Son of God bowed his head and died. The price had been paid. Nothing more was needed. Nothing more could add to the payment that Jesus made. It was finished.

When Jesus died, the Bible records that darkness suddenly covered the land in the afternoon, and there was an earthquake. The tough Roman soldier who stood at the foot of the cross watched all these things and cried out *"surely this man was the Son of God"* (Matthew chapter 27, verse 54).

Unfortunately, religions and even offshoots of Christianity have tried to say what Jesus did was not enough and we must do more to pay for our sin and separation from God. They have missed the point entirely. We could never pay for our sin. If we could, then Jesus did not need to die on our behalf. People add to it because they feel it's unfair or not right that we don't have to pay for our sin. But that's the whole point. We never could.

The Son of God, dead on a cross. If the story ends there then it's a very sad story indeed. If death had victory over the Son of God, then what hope is there for us?

We not only needed someone who could pay the price of our sin and separation from God, but we also needed someone who was able to beat the ultimate enemy of ours which is death. Remember, God said to Adam and Eve that because of their sin they would die both physically and spiritually. We needed a saviour who could conquer both on our behalf. The obvious questions then are did he die, and more importantly did he rise from the dead?

The Roman soldiers made certain Jesus was dead, and he was then taken off the cross and placed in a tomb with a large stone rolled across the entrance. The religious leaders were concerned Jesus' followers might steal his body and pretend he rose from the dead, so with Pilate's approval, they placed Roman guards at the tomb to ensure no one would steal the body. These professional soldiers made sure they did not fail. Failure would likely have come at the cost of their lives.

On the third day after Jesus died, the soldiers were standing by the tomb and an earthquake occurred. The soldiers saw an angel appear and roll the large stone back and sit on it. Not surprisingly, the Bible records the soldiers were terrified.

When Jesus' followers came to visit the tomb, they found the stone rolled away and the tomb empty. An angel appeared to them and said *"Do not be afraid, for I know that you are looking for Jesus who was crucified. He is not here; he has risen, just as he said he would."* (Matthew chapter 28, verse 5). Understandably, some of Jesus' followers who were not there were sceptical when they heard this account and ran to the tomb to see for themselves. They left the empty tomb wondering what had happened.

Over the next 40 days, Jesus appeared to over 500 people at different times and places. On various occasions Jesus appeared to groups of them at the same time and let them touch him to make sure he was not a ghost. He spoke to them, and even ate food in their presence to show he was for real.

Jesus didn't remain on earth. Instead, he gave his followers a command to go and tell the world that the price of sin had been paid and that all mankind could have their sins forgiven by putting their trust in him.

Jesus' followers took this message and spread it as far and wide as they could. Each of them had personally seen the risen Jesus and they were willing to die for what they had seen.

———◆●◆———

Now, at this point, I hear you call time out. There's an important and obvious question that must be addressed, namely how do we know Jesus rose from the dead?

The resurrection of Jesus is the most important component of the Christian faith. If Jesus did not rise from the dead, then Christianity is false. The entire message of the Bible hinges on the belief that God sent his perfect son to die on our behalf, to conquer death, and to offer us the gift of eternal life.

If Jesus truly rose from the dead, it changes everything. It means there is an afterlife and affirms that what Jesus said about himself and God is true.

However, if Jesus did not rise from the dead, then we are left with a man who people thought was a good moral teacher and yet he lied about his own resurrection. The Bible even calls out the importance of the resurrection. It says *"...if Christ has not been raised, our preaching is useless and so is your faith."* and *"...If only in this life we have hope in Christ, we are of all people most to be pitied"* (1 Corinthians chapter 15, verses 14-19). Even those who wrote the Bible were honest enough to say that the whole thing is a load of nonsense if Jesus did not rise from the dead. That's great because it puts the claim of the resurrection under the microscope. It sounds like a simple one to test and discount and many people have taken on this challenge.

The resurrection of Jesus has been vigorously tested by sceptics from a range of backgrounds including lawyers, scientists, investigative journalists, and philosophers, and yet the truth of the resurrection has withstood every argument against it.

It is beyond the scope of this book to provide a detailed analysis of this topic and I would strongly recommend reading *'The case for Christ'*, by Lee Strobel, an investigative journalist. As an atheist, Lee was disappointed his wife became a Christian, so he decided to investigate and put the issue of Christ's resurrection to bed once and for all. To his surprise, his journey convinced him that Jesus did rise from the dead and he became a Christian.

As a sceptical law student, Josh McDowell was also challenged to examine the claims of Jesus Christ. Initially, he considered the claims would not stand up under intellectual scrutiny, but after a detailed analysis he concluded the claims of Christ were true and he became a Christian. The results of his examination are detailed in his book *'Evidence that Demands a Verdict'*.

How can we examine this extraordinary claim of resurrection? Given resurrection is not an everyday occurrence, the threshold for proving it happened should be very high, and rightly so. A logical approach to test such a claim is to throw the darts of objection and see where they land. For our purposes, we briefly address the most common objections below.

Jesus didn't really die on the cross

A common objection to the resurrection is that Jesus never died in the first place. He was taken off the cross, no doubt in a serious situation, but he soon recovered and appeared to his disciples before leaving Jerusalem to avoid any further trouble.

Given resurrection is an extraordinary claim, it is understandable that a logical explanation is Jesus never died in the first place. Even the Islamic book, the Qur'an states that Jesus was not crucified and did not die. [64] However, under scrutiny, this objection doesn't hold any weight.

The theory of Jesus not actually dying was raised more than 200 years ago and has since been thoroughly examined by medical experts and pathologists and has been the topic of medical journals.[65]

In his book 'The Crucifixion of Jesus', Forensic Pathologist Frederick Zugibe notes that the theory Jesus never died is completely unfounded and contradicted by medical evidence. In addition to noting that Jesus could not have survived the crucifixion, he also notes there were no drugs or medicine available in Jesus' day to stop the pain he endured or to put him to sleep in a bid to make him fake his death.

In the Journal of the American Medical Association, a peer reviewed paper titled 'On the Physical death of Jesus Christ', the authors noted *"Clearly, the weight of the historical and medical evidence indicates that Jesus was dead before the wound to his side was inflicted and supports the traditional view that the spear, thrust between his right ribs, probably perforated not only the right lung but also the pericardium and heart and thereby ensured death. Accordingly, interpretations based on the assumption that Jesus did not die on the cross appear to be at odds with modern medical knowledge"* [66]

When the tomb was discovered empty, those who orchestrated his crucifixion never questioned whether he died. They had witnessed his death and were satisfied that the job was done. For his enemies, the question was not whether he had died but rather why the tomb was empty.

The eyewitness accounts, the historical accounts, and the medical and pathological analysis are all in agreement that Jesus most definitely died on the cross.

Jesus' body was stolen from the tomb

At first glance, the idea that Jesus' body was stolen from the tomb is plausible. Some argue that it was Jesus' disciples who came to the tomb after Jesus was laid to rest and stole the body. They then spread the rumour that the tomb was empty. Interestingly, the Bible also records that this theory was put forward on the first day the tomb was found empty.

When Jesus died on the cross, a wealthy man named Joseph of Arimathea went to the Governor, Pontius Pilate, and asked for permission to take the body down from the cross. As a follower of Jesus, Joseph and another follower named Nicodemus took the body and laid it in Joseph's tomb.

The gospel of Matthew records that the religious leaders were concerned the disciples might steal the body and spread a lie about the resurrection, so they also went to Pilate and asked him to secure the tomb to avoid this from happening. Pilate ordered a group of soldiers to stand at the tomb and guard it from interference.

On the third day after Jesus' death, the guards were terrified to see an angel come and roll the stone away from the tomb. Matthew (chapter 28, verses 2-4) records this scene *"There was a violent earthquake, for an angel of the Lord came down from heaven and, going to the tomb, rolled back the stone and sat on it. His appearance was like lightning, and his clothes were white as snow. The guards were so afraid of him that they shook and became like dead men."*

The guards ran off and went into the city to tell the religious leaders what had happened. Matthew (chapter 28, verses 12-15) records the first conspiracy of the resurrection. The religious leaders *"devised a plan"* and gave the guards a large sum of money, instructing them to say, *"His disciples came during the night and stole him away while we* were asleep." The rulers also told the guards "If this report gets to the governor, we will satisfy him and keep you out of trouble." Evidently, the guards spread the story because the book of Matthew records "...the soldiers took the money and did as they were instructed. And this story has been widely circulated among the Jews to this very day."

For several reasons, the idea of the body being stolen does not stand up under scrutiny. A group of professional soldiers with orders directly from Pontius Pilate, were specifically instructed to guard the tomb which they did. This was a highly volatile situation. The town was in the middle of celebrating a significant Jewish festival with thousands of visitors in attendance. Everyone knew of this famous teacher named Jesus, and his crucifixion had caused shock waves through the streets of Jerusalem. The last thing Pilate wanted was further trouble and the guards understood the importance of doing their job.

The idea that the guards would be asleep on duty is not plausible. However, even if they were, the idea they would continue sleeping while a large stone was rolled away from the tomb and a group of men carried off the body is even more implausible. Furthermore, if they were asleep, how would they know who stole the body?

In addition, the Bible records the burial cloths Jesus was wrapped in were neatly folded and left in the tomb. If the disciples managed to roll the stone away while the guards slept, they would certainly have been in a hurry to get out of there. They would not have been thinking about removing the cloths from the body and tidying up after them.

The guards knew that to fail in their orders would have had dire consequences. In the book of Acts (chapter 12, verse 19), we read a story of guards who were asleep when an angel rescued prisoners they had been instructed to guard. The next morning, the guards were cross-examined by Herod and then executed. This was serious business. The soldiers guarding Jesus' tomb certainly followed their orders. They knew the consequences, but on this occasion, they were outdone by an angel.

Even if the disciples stole the body from the tomb, it does not explain what drove them to ultimately die for their testimony that Jesus had risen from the dead. The disciples would have known the claim of resurrection was false, and yet they all went on to not only preach that he had been resurrected, but they were willing to be beaten, tortured, and ultimately killed for the message they spread. It's one thing to die for a lie you genuinely believe in, but it's another to die for a lie you know to be a lie because you invented it. Except for the Apostle John, these disciples did not merely die of old age. They died under extreme circumstances and pain. They all died alone and yet even being singled out, not one of them changed the story of what they had witnessed.

The disciples were hallucinating when they saw Jesus

The Bible records that many people claimed to have seen the risen Jesus on multiple occasions. In the book of 1 Corinthians (chapter 15, verse 6), we read *"...he was raised on the third day according to the Scripture, and that he appeared to Peter and then to the twelve. After that, he appeared to more than five hundred of the brothers at the same time, most of whom are still alive, though some have fallen asleep."* The author was making it clear that the readers of his letter to the Corinthians could check the story out for themselves because the eyewitnesses were still alive.

The book of Acts (chapter 1, verse 3) records that Jesus appeared to people for 40 days after his resurrection and many witness statements are recorded in the Bible.

Mary Magdalene was the first to see Jesus outside the empty tomb. After this, he appeared to his disciples while they were together in a locked room. The disciples were understandably terrified of the Jewish leaders and wondered if they would be next in line for the death penalty. In this locked room, Jesus suddenly stood among them. The disciples were startled. Some thought he was a ghost, and to demonstrate he was alive, he asked if they had any food and then ate it in front of them. On another occasion, Jesus appeared again to his disciples in a locked room and said to them *"Put your finger here; see my hands. Reach out your hand and put it into my side. Stop doubting and believe"* (John chapter 20, verse 27). I find that interesting. The Bible doesn't try to fluff up the story. It gives an honest account that even Jesus' closest disciples needed to be convinced that he was truly alive again. We need to put ourselves in their shoes. They had seen him die and now there he was in front of them. It's no wonder that Jesus continued to appear to them. They needed all doubt to be removed.

Many other eyewitness accounts are recorded, including two men who were walking outside of town and Jesus suddenly walked alongside them, explaining to them the prophecies of the Old Testament that spoke about him dying and rising to life.

Perhaps the most personal of appearances was Jesus appearing on a beach while the disciples were fishing and then calling them to come to shore and eat some food he had cooked for them on the beach. Afterward, he walked with Peter on the beach and encouraged him following Peter's denial of Jesus at the time of his death.

One of the greatest witnesses a lawyer can have in court is someone who used to be adamantly against their client but then completely changes their statement, especially when there is no reason to do so. The Apostle Paul was such a person. He was opposed to Christianity and personally arrested and killed Christians. However, most of the New Testament is written by him. How can that be?

The Bible records that on one of his journeys to arrest Christians, Paul and his companions were struck with a blinding light from heaven and the resurrected Jesus met Paul personally and spoke to him. As a result, Paul devoted his life to spread the news of what he had seen and the message of the resurrection, ultimately dying for his belief.

The historical record is clear. Many people claim to have seen the risen Jesus, not only on their own, but in groups. The evidence is further corroborated by the lives of those eyewitnesses after having seen him. They all maintained their eyewitness account right through until their deaths. This presents an enormous problem for the theory that people were having a hallucination. There is no evidence that group hallucinations occur. Psychologist Gary Sibcy wrote, *"I have surveyed the professional literature (peer-reviewed journal articles and books) written by psychologists, psychiatrists, and other relevant healthcare professionals during the past two decades and have yet to find a single documented case of a group hallucination..."* [67]

The hallucination theory does not stand up. In addition, hallucinations do not account for the empty grave.

People looked in the wrong tomb

The idea with this objection is that the disciples went to the wrong tomb and finding it empty presumed that Jesus had risen from the dead and spread the news. However, this theory falls apart very quickly.

The Jewish religious leaders and the Romans knew which tomb Jesus was placed in. Joseph of Arimathea clearly identified the tomb as his own personal tomb. The religious leaders could have easily put the entire issue to bed by merely showing people the correct tomb and the dead body.

Furthermore, Matthew (chapter 27, verse 61) records that as Joseph and Nicodemus laid Jesus in the tomb *"Mary Magdalene and the other Mary were sitting there opposite the tomb."* They knew which tomb it was, and they were the first to discover it empty.

There is no avoiding the fact that at the time, everyone on both sides of the issue agreed that Jesus died, that he was buried in Joseph's tomb, and that on the third day, Joseph's tomb was empty.

———————•♦•———————

What if it's all true?

Using the standards of evidence our courts rely on, all objections to the resurrection of Jesus fall apart.

Granted, Jesus rising from the dead is an extraordinary claim. However, the evidence is equally extraordinary. Let's stand back and consider what evidence we have to lean on.

Jesus was indisputably a good man who predicted his death and resurrection on multiple occasions and verified his authority and authenticity with a range of miracles. His enemies plotted against him, crucified him, and ensured he was dead. His place of death was known by everyone and was heavily protected by professional guards. On the very day Jesus predicted, his tomb was found empty and numerous witnesses including the guards, confirmed an angel was present at the time and was responsible for rolling the stone away from the tomb. From the moment his tomb was discovered empty and up until 40 days later, over 500 people confirmed having had personal encounters and conversations with Jesus, and many of his disciples went on to not only testify to what they saw, but also willingly die for their belief.

If the resurrection truly happened, it is the greatest message the world has ever heard because it means Jesus has conquered death on our behalf and opened a way for us to know God again. The question then is what does it mean for me personally? What must I do? How can I know God?

The Bible answers that question in these words *"For God so loved the world, that he gave his one and only Son, that whoever believes in him shall not perish but have eternal life. For God did not send his Son into the world to condemn the world, but to save the world through him. Whoever believes in him is not condemned, but whoever does not believe stands condemned already because he has not believed in the name of God's one* and only Son." (John chapter 3, verses 16-17)

There's nothing you can do to make things right with God. You can't be good enough, and you can't pay the penalty for your sin. But God in his infinite love, extends his hand and offers you a gift that he has paid for and asks you to receive it. How do we do that?

The book of Romans (chapter 10, verses 9-10) says there are two things we need to do to receive the amazing gift from God. It says *"if you confess with your mouth 'Jesus is Lord' and believe in your heart that God raised him from the dead, you will be saved."*

The first thing we must do is confess Jesus as Lord. What does that mean? Jesus is Lord of everything. He is the creator, and he is the King of kings. When we confess Jesus as Lord, it means we are not only acknowledging who he is, but we are also declaring that we are placing our life in his hands and asking him to be the Lord of our lives. That makes it personal.

When we confess Jesus as our Lord, we get off the throne of our life and let Jesus take his place. He becomes our king. He sits in the driver's seat.

Jesus explained what this means in the book of Matthew (chapter

16, verses 24-25) *"...If anyone would come after me, he must deny himself and take up his cross and follow me. For whoever wants to save his life will lose it, but whoever loses his life for me will find it"*. That is a statement of being all in. If you lived in Roman times and saw someone carrying a cross, you knew they were as good as dead. They were not making any plans for the weekend. It's the same with us. If you take up your cross it means you are dead to yourself.

Confessing Jesus as Lord means a complete surrender of our lives and desires to him. It means that instead of living for ourselves, we turn from all our sin and lay our lives at the throne of Jesus and say, *'you are now my Lord'*.

———— ◆●● ————

The second thing we need to do is *"believe in your heart that God raised him from the dead"*. What does this mean and why is it important? It means we acknowledge that instead of us paying for our sin, Jesus did it for us when he died on the cross. He paid the price we could not pay. In addition, we believe the payment for our sin and separation from God is full and complete, and the proof is that God raised Jesus from the dead.

The book of Ephesians (chapter 1, verse 14) says that when you believe in Jesus as your Lord and saviour, he does something to seal the deal *"...Having believed, you were marked in him with a seal, the promised Holy Spirit, who is a deposit guaranteeing our inheritance..."* When we believe in Jesus, God places his Holy Spirit inside us. Our relationship with God is restored and we live the remainder of our lives getting to know him more until we enter Heaven to live eternally with him.

———— ◆●● ————

What a journey we have been on to arrive at this message that God has for us all. It's a message for the whole world, but more importantly, it's a message for you personally.

The message is that God knows you like no one else does and loves you like no one else can. He came to earth especially for you. He paid the price for your sin, and he now extends a personal invitation for you to receive the free gift of eternal life and a restored relationship with him.

I'd also like to extend that invitation to you today, right now as you finish reading this book. Are you ready to ask Jesus to be the Lord of your life? If you are ready to make that commitment to him, then I invite you to have a conversation with God wherever you are. Maybe you haven't had one before. It's not hard. God is listening and when you talk to him, he hears your prayer.

You might want to say something like this,

"Dear God, I thank you that you sent your son Jesus to this world. I believe he died on the cross for my sin and that you rose him from the dead.

I ask that you forgive me for my sin, and I turn away from the life I have been living.

I give my life to you and declare that Jesus is now Lord of my life. I ask that you come and live within me. Amen"

———◆●◆———

If you have prayed that prayer, then I welcome you into the new life you have just commenced. If you know someone who knows Jesus already, then let them know, and ask them about how to join up with others who have done the same.

Remember, this is not a religion you have just joined, it's a living relationship with God. His Holy Spirit has come to live within you. You can get to know him more through reading the Bible, and talking to him. If you have a Bible, a great place to start is reading the book of John that tells you about the life of Jesus.

———◆●◆———

We started this book talking about purpose. It is my hope that in reading this book, you have not only concluded there is a God, but you have come to believe God loves you, and that you have asked him into your life. This is the ultimate purpose for your life. In the book of John (chapter 10, verse 10), Jesus said *"I have come that you might have life, and have it in abundance."* Jesus came to give you not only eternal life, but a life right now that is full of purpose, joy, and fulfilment. I encourage you to grab hold of it.

I would love to hear how your journey is going. Drop me a line at **mark@reesthomas.co.nz**

ABOUT THE AUTHOR

Mark Rees-Thomas was born in Wellington, New Zealand and educated at Victoria University of Wellington with a Bachelor of Law and Bachelor of Commerce. He has practiced Law since 1996.

Mark became a Christian as a young man, and after examining and verifying the claims of the Bible, he has continued to teach others about the message of Jesus.

Mark is married to Janine, and they have two children, Harrison and Isabella, and one son-in-law Hamish

END NOTES AND REFERENCES

1. An Introduction to Metaphysics (1935), Martin Heidegger

2. https://www.jmtour.com/personal-topics/evolution-creation/

3. https://www.britannica.com/topic/Ramapithecus

4. https://evolutionnews.org/2013/06/the_fall_of_aus/

5. https://creation.com/more-evidence-australopithecus-an-extinct-ape

6. https://creation.com/homo-erectus-to-modern-man-evolution-or-human-variability

7. https://creation.com/homo-erectus-to-modern-man-evolution-or-human-variability

8. https://en.wikipedia.org/wiki/Piltdown_Man

9. https://en.wikipedia.org/wiki/Nebraska_Man

10. The Language of the Genes, Revised Edition, London, Harper Collins, 2000, pg35.

11. https://www.passbiology.co.nz/biology-level-3/human-evolution#h.p_ID_54

12. https://www.keepinspiring.me/aristotle-quotes/

13. https://www.goodreads.com/author/quotes/275648.Socrates?page=9

14. Mere Christianity, C.S. Lewis, Harper Collins, pg 135-17.

15. Letter 12041, Darwin, C.R to Fordyce, John, 7 May 1879, Darwin Correspondence Project.

16. Quoted at Wikipedia, 'Galileo Affair', https://en.wikipedia.org/wiki/Galileo_affair

17. Beyond Natural Selection, Cambridge, MIT Press, Robert Wesson, 1991 p206.

18. Cited in 'Has Science buried God', John Lennox, pg 115.

19. Noted in 'Has Science buried God', John Lennox, pg 114.

20. David Gelernter, Claremont Review of Books, Volume XIX, Number 2, Spring 2019.

21. Interview on 23 July 2019 titled 'Mathematical Challenges to Darwin's Theory of Evolution', Hoover Institution.

22. The Wall Street Journal, Dec 24, 1997, article by Jim Holt, 'Science Resurrects God.'

23. Author unknown. Referenced at https://www.sunnyskyz.com/funny-jokes/20/Sherlock-Holmes-and-Dr-Watson-Go-Camping

24. Has Science buried God, John Lennox, Pg 175.

25. The Concise Oxford Dictionary, Clarendon Press, Oxford, 1990, eighth edition.

26. Some research suggests that chimpanzees have a limited ability to do so.

27. https://en.wikipedia.org/wiki/Cogito,_ergo_sum

28. https://www.vox.com/future-perfect/2023/6/30/23778870/consciousness-brain-mind-hard-problem-neuroscience-koch-chalmers

29. https://www.nature.com/articles/d41586-019-02207-1

30. ABC Television 20/20, 1989.

31. New York Times, 12 March 1991, P. B9.

32. https://news.utexas.edu/2023/04/13/james-webb-space-telescope-images-challenge-theories-of-how-universe-evolved/

33. Franklin Institute Journal, March 1936 'Physics and Reality'.

34. Has Science Buried God, Pg 62.

35. https://blog.fold3.com/april-4-1945-the-liberation-of-ohrdruf/

36. https://blog.fold3.com/april-4-1945-the-liberation-of-ohrdruf/

37. https://newspapers.ushmm.org/events/eisenhower-asks-congress-and-press-to-witness-nazi-horrors

38. https://www.shapell.org/manuscript/general-eisenhower-ohrdruf-concentration-camp/#transcripts

39. https://arxiv.org/pdf/1602.00690v1.pdf

40. Reference is made to 'Has Science buried God', John Lennox; and 'Big Bang Refined by Fire', Dr Hugh. Ross 1998, 'Reasons to believe', Pasadena. CA, and 'The Privileged Planet'

41. Annual Reviews of Astronomy and Astrophysics, 20, 1982, p.16.

42. God and the new Physics, London, J.M. Dent and Sons, 1983

43. https://www.popsci.com/earth-spin-faster/

44. https://evidencetobelieve.net/fine-tuning-of-the-universe-2/

45. The Privileged Planet pg 58.

46. The Privileged Planet pg 107.

47. https://imagine.gsfc.nasa.gov/science/objects/milkyway

48. https://arxiv.org/pdf/1112.4647.pdf Luke A Barnes, Institute for Astronomy ETH Zurich, Switzerland, 2012.

49. https://evolutionnews.org/2019/04/chemist-james-tour-is-scathing-hilarious-show-me-the-chemistry-of-abiogensis-its-not-there/

50. http://www.originthefilm.com/mathematics.php

51. http://www.originthefilm.com/mathematics.php

52. Undeniable: How Biology Confirms Our Intuition That Life Is Designed, Douglas Axe, July 12, 2016 by HarperOne.

53. https://www.jmtour.com/personal-topics/evolution-creation/

54. https://inference-review.com/article/animadversions-of-a-synthetic-chemist

55. Noted in 'Has Science buried God', John Lennox, pg 112.

56. https://www.drmingwang.com/

57. https://www2.drmingwang.com/amcl.html

58. https://david.dw-perspective.org.uk/da/index.php/writings/creation-and-mathematics/

59. Refer to https://www.icr.org/creation-logic for a detailed analysis of mathematics in creation

60. https://compassclassroom.com/fibonacci-numbers/#

61. Image from https://www.coolmathgames.com/blog/how-to-play-atari-breakout Code generated from ChatGPT

62. Mere Christianity, C.S. Lewis, Harper Collins, pg 135-17

63. The Authority and Inspiration of Scripture, Professor Van Til

64. https://quran.com/4/157

65. https://en.wikipedia.org/wiki/Swoon_hypothesis

66. Edwards, M.D., William; Gabel, M.Div., Wesley; Hosmer, M.S., Floyd (21 March 1986). 'On the Physical Death of Jesus Christ'. Journal of the American Medical Association. 255 (11): 1455–63. CiteSeerX 10.1.1.621.365. doi:10.1001/jama.1986.03370110077025. PMID 3512867

67. The Resurrection of Jesus: A New Historiographical Approach (Downers Grove, IL: IVP Academic, 2010), p.484.

All Bible references unless otherwise stated are from the
New International Version 1984

Also by Mark Rees-Thomas

The Tale of Christmas

Available online through:

Manna Christian Stores
www.manna.co.nz

Amazon
www.amazon.com

www.ingramcontent.com/pod-product-compliance
Lightning Source LLC
Chambersburg PA
CBHW070933210326
41520CB00021B/6928